"Heinrich, who combines his keen scientific eye with the soul of a poet, enthralls with this new, captivating and at times surprising examination of animal survival in the coldest of seasons."

—*New York Times Book Review*

"The cleverness of evolutionary design is everywhere on display in this look at how animals cope with winter.... The stories are plain engrossing—in their elucidation, their breadth of examples, and their barely contained sense of awe and admiration."

—*Kirkus Reviews*

"Critics and prize-givers have long raved about Bernd Heinrich's nature writing, but for those unacquainted with the work of this Maine woodsman, his new book, *Winter World: The Ingenuity of Animal Survival*, is a perfect introduction."

—*Minneapolis Star Tribune*

"It's like snowshoeing with Gary Snyder, tromping through the forest as someone points out a golden-crowned kinglet and explains how the tiny birds defy the odds and the laws of physics, and prove that the fabulous is possible."

—*Outside* magazine

"Anyone who has ever pulled on long underwear or Sorel boots will be captivated by what Heinrich reveals about how some birds, mammals, reptiles, and insects make it through the frigid season. ... Heinrich presents scientific explanation with great clarity, but his prose makes for good reading because he links fact with enthusiasm and experience."

—*Pittsburgh Post-Gazette*

BERND HEINRICH

ecco

An Imprint of HarperCollins*Publishers*

NEW YORK • LONDON • TORONTO • SYDNEY • NEW DELHI • AUCKLAND

WINTER WORLD

The Ingenuity of Animal Survival

A hardcover edition of this book was published in 2003 by Ecco, an imprint of HarperCollins Publishers.

P.S.™ is a trademark of HarperCollins Publishers.

WINTER WORLD. Copyright © 2003 by Bernd Heinrich. All rights reserved. Printed in the United States of America. No part of this book may be used or reproduced in any manner whatsoever without written permission except in the case of brief quotations embodied in critical articles and reviews. For information, address HarperCollins Publishers, 195 Broadway, New York, NY 10007.

HarperCollins books may be purchased for educational, business, or sales promotional use. For information, please e-mail the Special Markets Department at SPsales@harpercollins.com.

All line drawings are by BERND HEINRICH.
The temperature conversion table on page 8 is by JANE S. KIM.

FIRST ECCO PAPERBACK PUBLISHED 2004.

FIRST HARPER PERENNIAL EDITION PUBLISHED 2009.

BOOK DESIGN BY SHUBHANI SARKAR

Library of Congress Cataloging-in-Publication Data has been applied for.

ISBN 978-0-06-112907-0

23 24 25 26 27 LBC 20 19 18 17 16

ACKNOWLEDGMENTS

First of all I thank the keen observers. I thank those who have cared enough about nature to ask questions, explore, experiment, think, analyze, and draw cold-eyed conclusions from empirical evidence. It is they who have created the splendors of nature, some of which I have shamelessly borrowed to write about.

Created? Yes. Nature exists. But the wonders of nature dwell in the minds of sentient beings who are receptive to them. Pressure waves produced by the swoosh of a raven's wing or the light rays reflected from its burnished feathers are both physical manifestations. They can be measured, but they are neither *sounds* nor *colors* until their energy is transposed into action potentials in living neurons, and the action potentials are then transduced into *sensations* by the brain. Similarly, the splendor we can perceive in a golden-crowned kinglet's survival of a cold winter night or how a snapping turtle endures while sealed under the thick ice of a pond for six months does not exist until revealed by (and to) a receptive brain.

I once read somewhere that the findings of biology put a "barrier between humanity and nature." Perhaps the author felt, like many of us do, that science implies detachment. It does to me, but only as a filter that sifts out the splendid nuggets from chaos and those that are revealed from those merely imagined. Far from being a distancing, the science of biology is the opposite. It comes from an intense desire to get to know something intimately: you can't hope to get closeness with the real thing unless you know its contours.

I also read somewhere that Thoreau "stopped being a thinker" when he became a naturalist. I think that's getting it the wrong way round. You need facts to think with, and thinking about nature without facts is, really, feeling. Fiction is fiction, no matter how real one tries to make it seem.

ASIDE FROM ALL OF THE anonymously produced material that I used freely, I also thank the following for their open discussions, criticisms and/or comments that helped me sift out the real from the imagined: Ross T. Bell and Douglas Ferguson (identifying insects), Thomas D. Seeley and Rick Drutches (bees), C. William Kilpatrick (mammals), David S. Barrington (plants), Ellen Thaler and Charles R. Blem (kinglets), F. Daniel Vogt (deer mice), Brian M. Barnes (Arctic hibernators), Kenneth B. Storey (hibernation physiology of insects and frogs), Jack Duman, Olga Kukal, and Richard E. Lee, Jr. (insect hibernation), Gordon R. Ultsch and Carlos E. Crocker (turtle hibernation), and Lincoln B. Brower (monarchs).

Chapters 16 and 24 are adapted from articles appearing previously in *Natural History* magazine and portions of Chapter 5 were previously published in *Audubon*.

Kimberly Layfield and Louise O'Hare typed the manuscript, always quickly, efficiently, and without delay. I sincerely thank Daniel Halpern and Lisa Chase, my editors, whose interest and enthusiasm were always encouraging and whose numerous queries and suggestions were invaluable. I thank my wife, Rachel Smolker, for understanding.

And I dedicate this book to "Bart," George A. Bartholomew, for introducing and teaching me about the wonders of physiological ecology.

CONTENTS

When I was a teenage boy in western Maine, I read
the books of Jack London, books about a world of
rugged people and hardy animals at home in the
frozen woods of the north. Dreaming of that
world, I ventured out into the forest on snowshoes,
and if it was in the middle of a storm, all the better.
Deep in the forest I would dig a shallow pit in the
snow and using the papery bark peeled from a
nearby birch tree and dead twigs broken from a red
spruce, I'd start a crackling fire. The splendor of
sparks shooting up into the dark sky, the acrid
smoke rising through the falling snowflakes, and
hare or porcupine meat roasting on a stick over the
flames, all enhanced the winter romance. Warming
myself, I would think of London's "To Build a Fire,"
a story about how in the northern wilderness,
heat meant life. To one unfortunate newcomer in
the frozen Yukon in that story, the key to life was

keeping dry and having a match, but because of careless mistakes, he got his feet wet and his fire and life were extinguished.

The trouble with that newcomer, London wrote, was that "he was without imagination. He was quick and alert in the things of life, but only in the things. Not in the significances. Fifty degrees below zero meant eighty-odd degrees of frost—[it] was to him just precisely fifty degrees below zero. That there should be anything more to it than that was a thought that never entered his head." The newcomer, the cheechako, knew about the abstract thing, frost, and the numbers. But he did not yet know what they *meant*. And with good reason, too, as we're adapted to a tropical environment and maintain it around ourselves all year long, through our housing and our clothing. Most of us already feel uncomfortable experiencing 32°F (or 0°C), the temperature at which water turns into ice. What would we know of -50°F? We don't experience such temperatures, so we can hardly imagine how animals survive; by the time the winter world descends, most of us have surrounded ourselves in an artificial tropics.

In my Jack London–obsessed adolescent self, I may have occasionally experienced chilling, but it was not sufficient to grab my attention. I was focused on making each of my outings into the winter woods an adventure. I remember creeping out of my bed with two friends at school one midnight to ski in the milky moonlight through the pine and hemlock woods by Martin Stream, in western Maine. In our minds we were on the Dawson Trail in the Yukon, where we had to be tough. After all, anything could happen at the edge. In our imaginations we could fairly hear the huskies' breath, and the dogs barking at a distant farm seemed like howling wolves. Just as London said, the northern lights flickered under greenish-purple curtains draping the heavens, giving legitimacy to our fantasies. A barred owl hooted nearby in the dark cedar swamp, snowshoe hares crisscrossed the balsam fir thickets in utter silence, and deer went on their errands over the ridges. Yet we were unconscious of the many other worlds of these creatures, and scarcely thought of how animals survived temperature extremes ranging down to -50°C. Like the cheechako on his first winter, I was without imagination because I was without experience.

Each species experiences the world differently, and many species have capacities that are far different from ours. They can show us the *unimaginable*. Thus, the greater our empathy with a variety of animals, the more we can learn. For example, nobody on their own would seek to collect a fluid that is practically indistinguishable from water out of a specific kind of tree, then evaporate it to produce sugar. But the Iroquois, the native people of New York State, say that maple syrup was discovered by a boy who noticed a squirrel licking some maple sap at a wound where the water had evaporated. He had discovered a squirrel's winter food. Curious, the boy tried the sap himself. Finding it sweet, he began the tribe's use of a new resource. Similarly, prior to actually observing by eavesdropping with sophisticated electronic equipment, nobody would have suspected that bats see the world with their ears, that elephant seals dive down to a mile in depth and can stay down an hour, that moths can smell a mate from a mile away, or that birds fly nonstop across oceans.

One of the early theories (by nineteenth-century German biologist Carl Bergmann) that relates to the winter world, which is now enshrined as Bergmann's Rule, is that northern animals are larger in size than their southern congeners, to allow them to better conserve the body heat that is generally costly to produce. It is therefore perhaps not surprising that the world's largest of the Passeriformes (the most common and species-rich group of birds), the common raven (*Corvus corax*), is a northern bird, and the largest individuals of this species live from Maine to Alaska. But as with most rules, Bergmann's applies only if everything else is equal. It never is; the world's *smallest* perching bird shares the raven's northern range, even in winter, and it weighs as little as $1/325$ the weight of a large raven. This northern companion to ravens, the golden-crowned kinglet (*Regulus satrapa*) weighs in at about 5 grams—the same as two pennies. It is scarcely larger than a ruby-throated hummingbird or a pygmy shrew, yet it appears to thrive in the northern winter woods. I saw these tiny birds on my boyhood excursions into the winter woods of Maine—and I see them now, and am still amazed when I step out in the morning after a cold night and I'm greeted by them. Our fragility in the cold makes the survival of these tiny creatures all the more miraculous.

Golden-crowned kinglets.

Sigund F. Olson wrote in *Reflections from the North Country:* "If I knew all there is to know about a golden arctic poppy growing on a rocky ledge in the Far North, I would know the whole story of evolution and creation." He could have substituted the kinglet for the poppy. Kinglets are drab-colored birds with a flaming red, yellow, or orange crest. When excited, kinglets can suddenly flash their bright crest out of their olive-colored head feathers. They are one of the most common yet least-known forest birds living in the Northern Hemisphere. When I see a kinglet hopping through a densely branched spruce tree covered with pillows of snow, I often imagine myself in its place, wondering how it experiences its world. Having a circumference of about the size of a walnut, the rate of heat flow from the body is increased over a hundredfold from what it is in my human state. The world is suddenly that much colder, and a fate of freezing to death in the northern winter becomes an almost nightly possibility. However, the wonder and the marvel of how kinglets survive cannot be understood or appreciated except when viewed through the window of the adaptations found in the numerous other animals that share its winter world. It is their special means of coping that form context and continuity for the mystery of how kinglets survive subzero temperatures. Each species opens, as Edward O. Wilson has said (in *The Future of Life*), "the gate to the paradisiacal world" that is "a wellspring of hope." I agree: If kinglets can do it, then anything seems possible.

Only if I knew how and why a golden-crowned kinglet survives a Maine or an Alaskan winter would I understand the story of winter survival. Like other animals of the north, its life is played out on the anvil of ice and under the hammer of deprivation. For those that endure until spring, existence is reduced to its elegant essentials. The kinglet is thus iconic not only of winter, but also of adaptability under adverse conditions. This bird symbolizes astounding and ingenious strategies that animals have evolved for coping in the winter world. It is emblematic of the winter world that I will here explore, since its diminutive size and its presumed diet of insects, when insects are hidden and in hibernation, combined to produce an unsolved mystery. For me, it was the kinglet that led me further and further into the winter world of the north woods, and into this book, spurring me on to find the miraculous.

Terms help and sometimes almost define what
we think. However, as much as possible I have
tried in this book to let empirical reality be the
ultimate arbiter, with terms serving only as handy
abstractions to encapsulate concepts. Unfortunately, concepts change and keep changing as new
information becomes available, so that the terms
should change as well. Throughout this book I
have used terms that have had various meanings at
different times and to different people. To minimize potential confusion and partly as a review, I
here try to clarify some of these terms that refer to
the winter adaptations of animals.

In *Winter World*, I primarily use the Celsius scale
to measure temperature. For weights and lengths, I
use the U.S. as well as the metric system. For those
readers who need to brush up on converting
between Celsius and Fahrenheit or

between U.S. and metric measures, here are a few quick formulas: 1
ounce equals 28.35 grams, and 1 inch equals 2.54 centimeters.

Since so much of *Winter World* is about temperature, fuller Celsius
to Fahrenheit conversion details are given in the scale that follows:

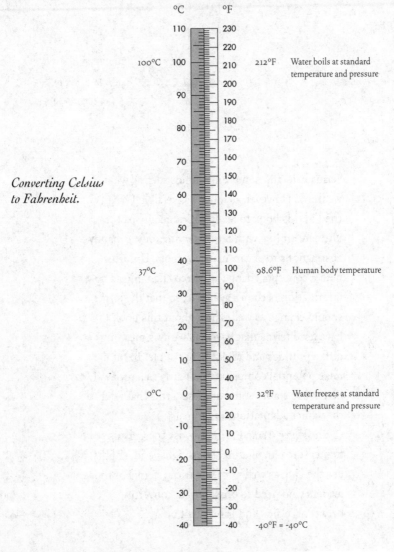

*Converting Celsius
to Fahrenheit.*

°C   °F

100°C   100   210   212°F   Water boils at standard
                                   temperature and pressure

37°C   37   100   98.6°F   Human body temperature

0°C   0   30   32°F   Water freezes at standard
                                    temperature and pressure

-40   -40   -40°F = -40°C

Most of the other terms in the book relate to hibernation, and even this term has caused confusion because of assumptions associated with it. Traditionally, hibernation simply meant winter inactivity, and it thus applies equally to frogs that have buried themselves in the mud under the ice, some insects and other frogs that are frozen solid while above ground, bears lying in their dens while maintaining a high body temperature, or ground squirrels and bats spending most of the winter with a low body temperature but periodically warming themselves up to be active for a day or more.

Hibernating animals are most (but not necessarily all) of the time in torpor, a state of inactivity achieved primarily (but not exclusively) by a greatly lowered body temperature. Hibernation refers specifically to an evolved suite of adaptations to the winter season, whereas torpor can either be a pathological breakdown of temperature regulation, or an adaptive response for conserving energy. Its duration can be for hours, days, or months.

After it became known that winter torpor, by setting down the body's thermostat, can be adaptive for warm-blooded animals, then low body temperature became almost the defining characteristic of hibernation. Precisely the same mechanism of adaptive torpor was then observed in some animals that survive not only winter conditions, but also inhospitable seasonal conditions in the desert. In this new context the "hibernation" physiology of torpor was then defined as aestivation.

Initially, the strict definition that joined the mechanism of hibernation (or aestivation) to body temperature implied that only mammals (and potentially birds) hibernated. However, since other animals that never regulate a high body temperature also engage in adaptive winter inactivity, a new term had to be invented, or else the old one needed to be discarded. The solution was to invent still a fourth term, brumation. Coined in the 1970s, this term refers to winter sluggishness or torpor of presumably cold-blooded amphibians and reptiles.

Later still it became widely known that some mammals and some birds routinely enter torpor to conserve energy, not just seasonally but also on a daily basis in summer. The behavior/physiology of torpor could then no longer be retained as the defining characteristic of hibernation even in the warm-blooded animals. Finally, as ever more numerous and varied ways of surviving winter were discovered, an all-inclusive definition of hibernation has become out of reach.

Body temperature turned out to be an especially inappropriate criterion for defining hibernation because many insects, presumed to be "cold-blooded," were found to regulate at times the same or even higher body temperature than the majority of birds and mammals. Like birds and mammals that at times allow body temperature to decline, they shiver (simultaneously contract opposing muscles otherwise used for locomotion to produce heat but little movement) so that they can become capable of rapid movement, in this case flight. Other insects stay *active* without ever heating up, either by shivering or by basking (increasing body temperature by orienting to capture solar heat rather than by shivering), and a few are even active with a body temperature at or slightly below the freezing point of water.

Activity and body temperature, as relating to the winter world, cannot be understood without rudimentary knowledge of the physical properties of water, and concepts and terms such as freezing point depression, antifreeze, ice-nucleation sites, thermal hysteresis and supercooling that will crop up later in the text and that I here foreshadow. In general, the freezing and melting points of water are the same temperature. For *pure* water it is common knowledge that solid-liquid transition occurs at a point defined as $0°$ on the Celsius and $32°$ on the Fahrenheit scale. (I shall refer primarily to the international, Celsius scale.) Solutes in water lower the freezing point predictably; adding one mole (molecular weight, which is a specific *number* of molecules) of any substance to a liter of pure water, for example, lowers the freezing/melting point by $1.86°C$. Many animal adaptations to the low temperature in the winter world relate to physical "tricks" of altering the predicted freezing point, by exploiting other physical phenomena associated with the freezing/melting point of water. First, the freezing/melting point depression is not *always* strictly a function of the molar concentrations of the dissolved substance in the water. *Some* substances—those special ones we call "antifreeze"—interact with the water molecules and cause a freezing point depression (lowering of the freezing point) that is *greater* than that *predicted* by concentration alone. An even more important phenomenon that some animals (especially insects) exploit is the *separation* of the freezing from the melting points. This anomaly is called thermal hysteresis. When a solution of water (regardless of whether it is pure or has solutes that may or may not be antifreezes) is in the liquid state when at a temperature below its predicted freezing point (i.e., in thermal hysteresis), then it is defined as being *supercooled*. Normally ice crystals form on and around some molecule or other ice crystal, and supercooling of a liquid is possible only in the absence of so-called nucleation sites around which ice crystals grow. Adding a nucleation site—such as a single ice crystal or a dust particle—to a supercooled liquid results in it all "instantly" turning into ice; and since supercooled liquids are not in a physically stable state, they can potentially freeze at any moment.

Another (nonexclusive) term sometimes used for overwintering insects is diapause, which is, however, more strictly defined as an arrested state of development. All insects are in arrested development when they hibernate (in part because low temperature, if not also freezing, retards or stops biochemical processes unless special mechanisms are invoked to circumvent the cold), but they are not strictly in diapause unless they do *not* respond with resumed development as soon as they experience warming. Many (but by no means all) moths arrest their development in the pupal stage during late summer and fall when it is still warm, and then hibernate as diapausing pupae. Others, depending on the species, hibernate in the egg, larval, or adult stage. Special adaptations are required for arresting development and combine with other traits for withstanding the cold during overwintering. Diapause also occurs in the absence of hibernation. For example, some adult insects enter *reproductive* diapause in the summer when they migrate or search for host plants.

The muddle in terminology of what hibernation means can be avoided if hibernation is defined not in terms of body temperature or some other specific physiological or behavioral phenomenon characteristic of a given species, but in terms of its adaptive function. In most animals hibernation and/or aestivation are seasonal periods of adaptive torpor that allow the animal to survive regularly occurring famine. Cold, heat, and aridity are factors that exacerbate the seasonal famine that hibernation has addressed through the evolution of various adaptations.

Even more confusion of terminology could be avoided by realizing that making ever more precise or restrictive definitions does not generate greater precision in the understanding of any animal. Animals are dynamic. Each animal's choices fit in somewhere in a long continuum of almost anything that can be measured or imagined. Different terms may apply in any one animal in varying degree, depending on circumstances, but ultimately the species, and often the individual, fashion their own solutions to fit the situation or the occasion. We gain understanding not so much by lumping and defining, but by differentiating the specifics from the generalized features. The latter have a tendency to become enshrined as rules or laws that are ultimately statistically derived descriptive artifacts. But animals don't follow rules or easily allow us to pigeonhole them into convenient intellectual boxes. A "rule" is nothing more than a consistency of response that we have deduced animals exhibit because it serves their interests. Rules are the sum of decisions made by individuals. They are a result. The chaos, and the art, of nature remains.

Microscopic life *evolved some 3.5 billion*
years ago in the Precambrian period during the
first and longest chapter of life that covers about
90 percent of geological time. No one knows
exactly what the earth was like when microbial life
began, but we do know that at some time the earth
was a hot and hellish place with an atmosphere
that lacked oxygen. Early microbes, probably blue-
green algae or bacterialike organisms, invented
photosynthesis to harness sunlight as a source of
energy. They took carbon dioxide out of the air
as their food, and they generated oxygen as a waste
product that further transformed the atmosphere
and hence the climate. They developed DNA for
storing information, invented sex, which produced
variation for natural selection, and evolution took
off on its unending and largely unpredictable
course.

Molecular fingerprinting suggests that every life-form on earth today originated from the same bacterialike ancestor. That ancestor eventually led to the three main surviving branches of life, the archaea, bacteria, and the eukaryotes (the organisms made of cells with a nucleus that include algae, plants, fungi, and animals).

Remnants of the first ancient pre-oxygen-using life may still exist little-changed today. They are thought to be sulphur-consuming bacteria now living only in the few remaining places where the ancient and to us hellish conditions still remain. These habitats include hot springs and deep oceanic thermal vents where water at 300°C (that stays liquid there rather than turning to steam because it is under intense pressure in depths of some 3,600 meters) issues up from the ocean floor. One of the species living at the edge of these hot water vents is *Pyrolobus fumarii,* which can't grow unless heated to at least 90°C, and which it tolerates 113°C. As the earth cooled new environments became available and new single-celled and then multicelled organisms evolved from these or similar species to invade ever-new and cooler environments.

Some cells much later also escaped their ancestral conditions by invading other cells, finding that environment conducive for survival and adapting to it. Such initially parasitic organisms ultimately evolved into cooperative or symbiotic relationships with their hosts. Perhaps the most fateful of these eventually mutually beneficial associations occurred when some Precambrian green algae successfully grew inside other cells, to ultimately become chloroplasts, while their hosts then became green *plants*.

The ability to capture solar energy that ushered in the multicellular life and the fantastic diversity of life we see today was followed by or concurrent with one other critical parasitic-turned-symbiotic cellular invasion. The availability of oxygen from plants led to energy and oxygen-guzzling bacteria, and when some of these invaded other cells they became mitochondria and their hosts became animals.

Mitochondria are the cell's source of power or energy-use, and having mitochondria with *access* to oxygen allowed vastly greater rates of energy expenditure. It made the evolution of multicellular animals possible. One of the ultimate expressions of the high-energy way of life that is powered by the use of mitochondria is, of course, animals like the kinglets that maintain a liveness at an, to us, almost unimaginably high and sustained rate through a northern winter.

The metabolic fires generated by the mitochondria can be fanned to run on high, given the availability of much oxygen, or they may be turned down low. Life is the process that harnesses, and more importantly, *controls* that fire. It produces heat, and heat is often synonymous with life.

Temperature is, to us, a sensation measured on a scale of hot to cold. Physically, it is *molecular motion,* and we can measure it with a thermometer because the greater the motion of the molecules of a substance, say mercury, the farther apart they are spaced. We measure this molecular expansion as mercury (or some other liquid) in a column is displaced up a calibrated scale. The molecular motion, as such, is not life but a prerequisite for it.

Heat, on the other hand, is the *energy* that goes in or out of the system to change temperature. Some substances must absorb more energy (from the sun for example) before their molecules are set into motion, raising the temperature. One calorie is the unit of energy defined to raise one gram of water one degree Celsius. Substances, like rock, heat up with much less energy than that required to heat water. Again, energy is not life, but a prerequisite for it, and life is insatiable for it. What is truly miraculous, therefore, is that life continues and even thrives in winter, when the sun is low.

There is no upper limit of temperature. Within our solar system, the surface temperature of the sun is about 6,000°C; the center is about 3,000 times higher, or 18,000,000°C. The lower temperature limit in the universe, on the other hand, is finite. It's the point at which all molecular motion stops and the heat energy content is zero. That temperature precludes living, but from adaptations to the winter world that I will discuss, it need not destroy life. Life can, at least theoretically, persist on hold at the lowest temperature in the universe.

Our Celsius scale is defined as a 100-arbitrary-unit division of heat energy content of water, between when water molecules leave the crystal structure to become liquid (0°C) and 100°C when the liquid water boils at sea level. The zero energy content of matter, or lowest temperature limit in the universe, is defined as 0°K on the Kelvin scale and it corresponds to -273.15°C or -459.7° on the Fahrenheit scale. Since life as we know it is water-based, the *active* cellular life that most of us are familiar with is restricted to the very narrow temperature range between the freezing and boiling points of water (which vary somewhat depending on pressure and presence of dissolved solutes) where the controlled rates of energy use become possible. We are composed mostly of water, and when the water in our cells freezes, i.e., turns into ice, it shreds membranes and kills.

Water influences life as profoundly at the *ecological* level as it does at the cellular level. Every fall in the North Temperate Zone we can observe the ecological effects of the various physical properties of water. Most of the creatures of the earth experience water as a transparent liquid that runs downhill and that can only be contained by barriers. For part of the year some of us also see water as a white powdery matter that sticks to the trees and the sides of the hills and that makes the woods look like a fairyland. This substance can be stacked into piles, tunneled into, and made into dwellings for man and beast. It can accumulate and become so dense and deep that we can't walk through it. It can shut out the light to plants and may crush them. In northern areas, when the tilt of the earth is appropriate, it may collect over long periods of time to create glaciers that transform the landscape, grinding down mountains and valleys. With a difference of just one degree Celsius, or less, water also can become a clear, glasslike substance that seals over the surface of lakes and allows us to walk across them with impunity.

The fate of almost everything in the winter world is ultimately determined by the crystallization of water. In a matter of a few hours that crystallization can change the physical surface of the earth, and in the course of millions of years it has profoundly changed the physiological, morphological, and behavioral characterizations of all organisms that have to contend with that magic transformation of liquid to crystal.

Every fall the winter world creeps up gradually and inexorably onto those inhabiting the Northern Hemisphere. As it does, the nights get progressively longer and colder. Less energy from the sun reaches the ground. First the water in topsoil freezes to form a solid cover (unless it's already snow-covered). The swiftest-flowing streams and brooks are the last to freeze over because the cold air–water interface constantly mixes. The cold that causes the water to freeze comes from the air *just above* the water. The water is at least slightly warmer. When the water is stirred (as in swift-flowing streams), the surface doesn't cool down to 0°C so quickly.

One night the inevitable happens: the bodies of water freeze solid. The temperature drops enough for water molecules on marsh grass stems, twigs, and leaves floating at pond's edge to slow their molecular momentum enough that they drop into stable crystalline positions. The stems, twigs, and leaves then serve as nucleation sites for ice-crystal formation. Like billiard balls rolling into pockets, the water molecules lock into position, first indiscriminately on any object they encounter, then on other molecules that have come to rest, forming an ice lattice. What little energy these molecules had left is now released as heat, the heat of fusion, 76.7 calories for every gram of liquid water turned to ice. (This heat is not enough to cause any appreciable temperature rise in the pond or lake because it is so quickly absorbed by the much larger mass of water. However, the sudden freezing of a small droplet isolated from others often causes an appreciable "exotherm" of several degrees Celsius.)

The ice crystals being formed reach out like sharp fingers over the surface of the water. They meet, interlock, and by morning the whole pond may be covered with a transparent pane of ice that physically separates the water denizens from those on land. Quite overnight, one can literally walk on water, a capacity obviously less a function of supernatural abilities than of the physical properties of water at temperatures below 0°C.

There is something quite remarkable, simple, and yet profoundly important that happens when water turns to ice in a pond. Compare this with what transpires when water turns to ice in a cloud. In a cloud, the ice crystals fall because water and ice are heavier than air and the gas phase of water. However, water becomes *lighter* when it transforms from a liquid to a solid state. If this were otherwise, then ice crystals would sink as soon as they formed on the surface of a pond. Heat near the bottom of the water would at first continually melt the ice crystals coming down, but at some point temperatures near the bottom would reach 0°C and lower. The water would then freeze from the bottom up, rather than from the top down. The ecological consequence of this phenomenon would be that there would be no bodies of water in the north. Sunshine in the summer would melt only the upper layers of ice, and any aspiring body of water would soon become a huge permafrosted ice lens.

Another ecologically important aspect of the behavior of water is that its density changes with changing temperature. Cold water is denser than hot water, and so cold water sinks as hot rises. As is also true for air. But, in water, the change is not so uniform. Water becomes densest at 4°C. As a result, when lakes warm up in the springtime from 0° to 4°C, as the ice melts, the surface water sinks. This denser water displaces the colder bottom water and its nutrients, which then rise toward the surface and feed the life above.

Geologically, the earth has experienced regularly recurring ice ages that are dependent on an astronomical cycle of the earth's tilt (Imbrie and Imbrie 1979). This, the Milankovich cycle (named for its discoverer, Milutin Milankovich), is currently in a *cooling* period that began seven thousand years ago. But at the present time we are experiencing global warming instead, because the cooling effect of the astronomical cycle is being overridden by a human-induced climate change. The burning of fossil fuels produces carbon dioxide gas that is accumulating in the atmosphere at a greater rate than it is being absorbed by forest trees and other plants. The carbon dioxide acts as a thermal blanket, trapping solar heat. Unlike the astronomical cycle, which is gradual and permits evolutionary adaptations, this new phenomenon in the history of planet earth is sudden. It will affect kinglets, and us.

Sometime early in October the brilliant foliage comes to rest on the forest floor. And then one morning those leaves are encrusted with the white ice crystals we call hoar frost. A few weeks later the first snowflakes, the conglomerates of innumerable snow crystals formed in the air, may come spiraling down out of a darkening sky. Kids of all ages focus on the largest ones and make a game of maneuvering under them to try to catch them on their tongues.

Wilson Alwyn Bentley, or the "Snowflake Man" as he came to be known, also caught snow crystals on microscope slides. He lived on his family's homestead in the village of Jericho, Vermont, along with his brother Charles, their parents, their grandparents, and later Charles's wife and their children. Life on the farm revolved around the chores and seasons, and on February 9, 1880, on his fifteenth birthday, Wilson received an old microscope as a gift from his mother. It changed his life. "I found that snowflakes were miracles of beauty," he would say later. "Every crystal was a masterpiece of design, and no one design was ever repeated."

As a result of Bentley's writing and photography on the subject, every schoolchild is now taught that no two snowflakes are alike, although he pointed out that "it is not difficult to find two or more crystals that are nearly, if not the same, in outline." Snow crystals were to him a metaphor for earth's beauty. They were a "road to fairyland" so that "even a blizzard becomes a source of keenest enjoyment and satisfaction, to man, as it brings to him, from the dark, surging ocean of clouds, forms that thrill his eager soul with pleasure."

Prior to Bentley, scientists and naturalists had commented on the structure of snowflakes, marveling less at their variety than at their six-cornered shape. In 1610, Johannes Kepler (famous for many discoveries, including the planets' elliptical and not circular paths) is credited with first questioning why, whenever snow begins to fall, its initial formation invariably displays the shape of a six-cornered starlet. I'm not sure we have the answer now, but I assume that the six-arm configuration somehow relates to the most economical way that water molecules align when forming the crystal, when that crystal is free to grow in all directions in the air.

At the age of seventeen, Bentley merged the powers of his microscope with that of the newly developed camera, to realize his dream of capturing images of the snow crystal's beauty. In a near miracle his father consented to pay the one hundred dollars it cost to purchase the elements required to fashion a primitive camera. Bentley struggled for weeks, experimenting, before on January 15, 1885, he developed the world's first photo-micrograph of a snow crystal, in the woodshed of the family farm.

Bentley eventually needed to share his photographs with someone who might appreciate them, so he traveled ten miles down the road from his farmhouse to the University of Vermont in Burlington to see Professor George Henry Perkins, a biologist, ecologist, and longtime teacher there. Professor Perkins was amazed at the quality of Bentley's work and told him that he absolutely had to write about it and show his snowflakes to the world. Bentley went back home and tried to write, but gave up in frustration. He went back to Perkins, appealed to him to put words to his photographs, and in 1898 an article by W. A. Bentley and G. H. Perkins titled "A Study in Snow Crystals" appeared in *Appleton's Popular Science Monthly*. Perkins, not only a scholar but also a gentleman, wrote that although he put the pages together from Bentley's notes and photographs, the "facts, theories, and illustrations are entirely due to [Bentley's] writing and enthusiastic study."

The article launched Bentley's lifelong career studying snow crystals; it apparently unlocked his writer's block as well. He went on to write fifty popular and technical pieces on snow, culminating with a book, *Snow Crystals*, in 1931, the year of his death, in which he published more than 2,500 of his 5,000-plus photographs.

A snow crystal is only the beginning of what makes snow. Snow *flakes* are composite masses of often hundreds of snow crystals that have collided on their long journey down from the clouds. The final size of a snowflake before it comes to rest depends on various factors including the numbers of crystals issued per cloud, the distance traveled, and the temperature. Snow falling early in the winter usually forms the largest flakes. Later in the season, when temperatures are lower, the ice crystals spawned by the clouds adhere to one another less readily. These crystals in colder air are brittle and constant collisions on the way down degrades them or smashes their intricate and beautiful structures. Arms of the crystals break off, and these tiny ice spicules then make up the snow. Driven by the wind, the ever more degraded crystal fragments are then packed into a tightly interlocking lattice on the ground that at low enough temperature, near -30°C, has the texture and appearance of Styrofoam. Indeed, walking on such snow at -30°C or less has the feel, and produces the sound, of walking on Styrofoam. It is the building material that is carved into blocks and has been used for centuries to construct winter homes.

Packed snow has superb insulating properties. An igloo efficiently retains warmth generated by a small oil-lamp and human bodies, while it effectively blocks the infinite heat sink of the sky above. Wind cannot penetrate its walls, even as oxygen and carbon dioxide freely exchange through the snow and the entrance tunnel. The tunnel reduces air convection, such as from wind, and at the entrance to the igloo the Inuit create an air lock—raised area that reduces the inflow of the colder and heavier air that hugs the ground. Typically, this raised area is covered with caribou hides.

Ruffed grouse that has tunneled into the snow to sit out the night or a snowstorm.

Snow also provides shelter at night for many kinds of birds, ranging from Siberian tits, ptarmigan, and ruffed grouse that burrow in and create igloolike snow caves. The birds leave tangible evidence that they sometimes tarry a long time in these snow shelters; I have found over seventy fecal pellets within a single ptarmigan snow cave near Barrow, Alaska, and in the Maine woods I routinely find over thirty where a ruffed grouse has spent the night. Often these birds also stay the day in their shelter, because on snowy, cold days I have flushed them out from underfoot under the snow even at noon.

From all appearance, many northern birds are excited by snow, especially that of the first snowstorm of the year. Both ravens and ptarmigan then become visibly animated, rolling, sliding, and bathing in the fluff when it is not yet packed. Owls, crows, finches, tits, and kinglets also bathe in snow (Thaler 1982).

Snow has been such a constant feature of their environment that many northern animals have become well adapted to it and now depend on it. Perhaps none depends on snow more than the snowshoe hare. The size of this hare's tracks are all out of proportion to the animal's size. As a result of its low foot-loading, the hare can walk, hop, and run very near the top of the fluffiest snow. As a consequence, the more that snow accumulates throughout the winter, the more easily the hare can reach its food, the fresh twigs of small trees and brush. Thus, the twigs feed the hares, who are in turn reincarnated into fox, bobcat, lynx, fisher, weasel, great horned owl, goshawk, and red-tailed hawk. Yet despite the hares' rapid recycling into other lives, their populations persist because of tricks of individual survival coupled with a legendary reproductive potential. (A female hare may have four litters per year, with up to eight young per litter. The young are furred when born and almost ready to run and soon ready to reproduce.)

Regardless of how fast the hares reproduce, it would not take long for the predators to deplete them if it were not for their camouflage. The snowshoe, also called varying hare, changes from brown to a coat of pure white fur in winter. However, the more effective the camouflage is for one season, the less effective for another, and a hare's trick to survival requires getting the timing just right. It is hard to be exact in when it is best to become white or brown, because the coat change requires a month or more, and a snowstorm can transform a landscape in minutes. Proximally the hare's timing is determined by day length, but ultimately it must be dictated by the average time that the ground is snow-covered; the timing of the genetically fixed color change of locally adapted hares necessarily reflects the historic patterns of when there was snow cover because off-color hares are the first to be eaten and have their meat be reconverted into the next life, that of predators.

In the woods in western Maine the hares are almost all white by the end of November, the most usual time that there is continuous snow cover. However, in some years when the first snowfall is late, the hares show up for the whole duration of that lateness as if they had been marked in hunter's fluorescent orange. Within a few minutes after it snows, however, they become practically invisible. I doubt that a hare knows whether or not it is invisible because the totally white hares I've seen on brown background made no apparent effort to hide. Nevertheless, they could still have some change in behavior that compensates for their inability to accurately time the molt with snowfall events. For example, I routinely see almost no hares' tracks in the woods of western Maine after early snows in late October and early November, although their tracks are common in the same year and at the same sites in December. I had first suspected that the hares might migrate, until I happened to walk one November up on the ridges of Mount Bald near my camp. A hare could get up there in minutes. Here in the spruces near the top, I suddenly found many hare tracks, making me wonder if the molted animals move up into the hills where the snow comes earlier, and then later come down after the first snow falls in at the swamps, their preferred habitat.

In March the hares' white fluffy winter fur begins to drop out and is again replaced with the summer brown. Golden-crowned kinglets take advantage of this fortuitous timing of the hares' coat change to gather the cast fur for insulating their nests.

The hares' winter survival depends not only on ability to hide, but also to run when needed. Unlike many animals in winter they can and do stay lean, accumulating essentially no body fat, because food is almost always within reach and food energy need not be stored. Being lightweight and big-footed gives them an advantage for moving quickly on the snow. But sinking in even a little bit slows a runner down. Yet there is a limit to how big the feet can be before they hinder rather than enhance running speed, and snowshoe hares are probably already close to as light- and big-footed as they can get. They have, however, another behavioral trait that is apparent at a glance (the patterns of their tracks in winter woods): hares follow others' tracks and thus pack down the snow, making well-trodden highways. Hares traveling along these paths then clip the twigs along the way, and knowing the paths well, get the jump on any predator giving chase.

The snow can be an enemy, too. Small animals in the subnivian zone, that area in or under the snow, can at times be sealed in when the upper snow surface melts in the sun and then freezes at night into a solid crust. Grouse sometimes get trapped under the crust and become prey to foxes. Shrews that emerge onto the crust and do not quickly find a hole back down to safety, may be taken by a predator or simply freeze to death.

Polar bears have nothing to fear from the crust. They dig their dens into drifts of snow where they have their cubs, suckling them in warmth and safety, and hibernate for the six months of the arctic winter. My Winter Ecology students, like polar bears and Athabaskan hunters, also make temporary shelters in mounds of snow.

Every winter I take ten to thirteen students with me to my camp in the Maine woods, where we live in my homemade log cabin (two-story) that is without electricity but with a woodstove. We get water from snow we melt, or from a well at some distance. We bake our own bread and have been known to fry our own voles. We take long unstructured walks through the woods in the first week. In the second two weeks everyone settles on an independent research project that I guide. The hard part comes when everyone gets back to campus during the spring semester and analyzes their results and writes their scientific reports.

Building snow shelters is not one of the official projects. But we occasionally make them nevertheless. We start by heaping up a great pile of snow. A few hours after the snow has been heaped up, ice crystals interlock and combine to make a solid mass that can then be excavated to produce a snug and warm cave for overnighting.

Near the top of any snowpack, the snow gets denser as the crystals bond together. Meanwhile, close to the ground, where it is warmer than at the surface, water vapor from disintegrating snow crystals migrates upward and recondenses and freezes onto the upper snow pack crystals. In time, the growth of the upper ice gains at the expense of the lower, and a latticework of ice pillars and columns and extensive air spaces at ground level create the subnivian space that is, in a sense, a continuous snow cave inhabited by mice, voles, and shrews.

Within this space, temperatures are physically "regulated" within a degree or two of the freezing point of water, all winter long. Several factors are involved. First, as already mentioned, snow affords remarkable insulation and, even at -50°C, heat rising from the earth generally keeps the temperatures near the ground close to 0°C. When both ice and water are balanced at near 0°C in the subnivian zone, the temperature is stabilized since whenever heat is lost through the snowpack to cool this space slightly below 0°C, there is then a water-to-ice-crystal conversion, which releases heat. Similarly, whenever ice turns to water, the process requires or uses up heat. Thus, as long as both ice and water exist side by side, they constitute a thermostat keeping temperatures constant. Only the amounts of ice and water vary, depending on the amount of heat loss or input.

In New England, the subnivian zone is the home of voles (a type of short-tailed mouse): principally the meadow vole (*Microtus pennsylvanicus*) and the red-backed vole (*Clethrionomys gapperi*) as well as the masked shrew (*Sorex cinerius*), smoky shrew (*Sorex fumeus*), pygmy shrew (*Microsorex hoyi*), and short-tailed shrew (*Blarina brevicauda*). Every spring, right after the snow melts, or just as the last inch or two is melting, I see the labyrinth of the *Microtus* tunnels fully exposed on the surface of the ground. Also fully exposed lie the grass nests of these rodents, many of which will soon be occupied by bumblebee queens starting new colonies.

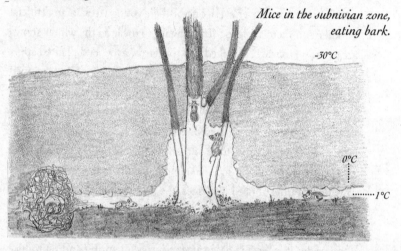

Mice in the subnivian zone, eating bark.

-30°C

0°C

1°C

The spring of 2001 provided an especially impressive demonstration of the importance of the subnivian world to meadow voles. Record amounts of snow had fallen in March in Vermont, and the voles appeared to be having a population explosion. Like lemmings, their close relatives, meadow voles have an awesome reproductive potential. One well-fed captive vole produced seventeen litters in one year, averaging five babies each after twenty-one-day gestation periods. The young females, in turn, can produce their own litters in one month. At such reproductive potential, it would not take long for them to carpet the earth. Luckily, such horrors of exponential growth are seldom realized. Instead, the voles' role in the economy of nature is, like that of hares, to convert vegetation into the protein-rich dietary staple of many predators that rely on them in winter, principally foxes, weasels, fishers, coyotes, and bobcats. The summer shift includes hawks and snakes.

In some areas all of the young sugar maple trees, box elder, and white ash trees were debarked right up to the snow line, but never above it. The tree damage caused by voles in winter is well known by orchardists, who would lose their young fruit trees every winter if they did not in the fall cover every one of their young trees in an artificial barklike commercial plastic stripping up to the level the winter snows are expected to reach. I learned this the hard way when I planted apple trees in the field by my cabin; by spring, every single one was stripped of bark for a foot above the ground, in what had been the subnivian zone in the winter. Older trees, once they have developed a thick layer of their own bark, are protected. The cambium, or inner live layer of bark of trees, is a favorite food of many herbivores, and the thick outer dead layer is essential armor. Like most armor, its usefulness is only apparent in time of need. For trees, the greatest need for thick bark is in the winter when more easy-to-eat foliage is not available.

Due to the protection of the snowpack and the cozy subnivian zone under it, voles are able to get a jump on spring and reproduce sometimes two to three months before the snowpack has melted. Wild spring flowers of many kinds also get an early start in the relative warmth of the subnivian zone under the snow. Some, like the snowdrops in our gardens, grow in March under the snow and send their flowers directly through the snow.

Peter Marchand, a winter ecologist who has extensively studied snow cover at the Center for Northern Studies in Vermont and elsewhere, has wondered how organisms that are buried under snow get their cue to start growing or breeding. How do they know, as they appear to, that the snowpack is about to melt off? Do they sense the sunlight? To investigate this problem, Marchand and his students studied the light-transmitting properties of the snowpack, finding that as the snow became increasingly more compact, it extinguished more and more light. But only up to a point. To their surprise, they found that when they mimicked the melting and refreezing that occurs in the spring when snow density increased, the snowpack became almost icelike in consistency. Then, despite or because of its greater density, it transmitted more light. Marchand speculates that this snow-penetrating light is sensed by the voles and stimulates them to start reproducing, thereby giving them their legendary reproductive potential. Alternatively, the plants detect the light first, and by growing, produce chemicals that then give the animals that eat them an indirect cue that then stimulates their reproductive activity.

For some animals in the winter woods, the subnivian zone is never totally separate from the subterranean zone. If it were fully separate, then few small mammals would survive the winter, because in some years subzero temperatures occur a month or two before there is an appreciable layer of snow. During these times, the shrews and voles inhabit the space under the leaf mold, or they live in rotting stumps that are riddled with cavities. At times they also burrow into the soil, living beneath the frost line. Still other animals, such as the molelike short-tailed shrew and the star-nosed mole and its cousin, the hairy-tailed mole, stay there permanently. The presence of the subnivian zone merely raises the frost level and allows them to be closer to the surface of the ground, where there are likely more insect prey.

While the snowpack is a haven for many small animals, it provides a severe challenge to the larger ones that hunt them for a living. Some predators would be unable to live in the north in winter if it were not for their specialized ways of hunting the subnivian prey. These winter-active hunters include the aforementioned weasels, foxes, and coyotes, and the great gray owl.

Foxes and coyotes locate mice by sound and pounce on them by crashing with their front paws through the snow. Their bouncing collapses the rodents' tunnels, temporarily trapping the intended victim. Great gray owls (*Strix nebulosa*) also have acute hearing and can detect a meadow vole's movements under snow from thirty meters away. Drawing near one, they plunge from twenty-five feet in the air, and with their balled-up feet can punch through crust thick enough for a person to walk on. They then catch the mice that are temporarily detained by the collapsed snow by repeatedly clenching their toes, sifting through the snow with their long talons. Great grays are among the largest of all owls, being nearly three feet tall, but a large part of their bulk is composed of thick layers of insulating feathers. They are not as powerful as great horned and snowy owls that specialize on hares. In calm weather the grays hunt both at night and during the day, and temperatures as low as -43°C do not cause them to leave their northern haunts. They do leave regularly, though, when their prey is depleted due to disease or overhunting. Rodent population crashes in one area do not preclude population explosions in another, and so the owls wander widely. So I too wander, a hunter of winter marvels.

Today, the sixteenth of March 2000, is the day
before St. Patrick's Day. It is still winter. In Vermont where I teach at the university, we are only
three-quarters through the woodpile, our stash of
stored solar energy collected by trees. Not a single
sprig of new green is yet in sight in all the woods;
the leaves, the most efficient and most beautiful
solar collectors ever devised, are not yet active. But
their construction material—the sugar to make
them—is already rising up the tree trunks, and in
only four more days the sun will cross the celestial
equator, and on this, the vernal equinox, the length
of the day will become equal to that of the night,
marking the official but not the real end of winter.
In only one more month the ice here will begin to
lose its grip. It has been a winter with much snow,
and I can't resist going out into the winter woods
to have a look around and experience winter. But I

must *choose* to experience in order to write, and I decide to try to focus, arbitrarily, to find the nest of a crossbill, a strange bird if there ever was one, in the woods of my camp in Maine.

Here is Wendell Taber's (in Bent 1968) experience with white-winged crossbills many years ago:

Smoke rises straight in the frosty stillness of an early September morning. Slowly the mist clears to reveal a tiny body of water. Tucked in at the 3,500-foot level in a region where the tree line is around 4,500 feet or less, Speck Pond lies nearly surrounded by the steep, towering, coniferous-clad walls of those wild Maine peaks, Mahoosuc and Old Speck. From across the lake comes a white-winged crossbill, then another, and yet another. Others appear, seemingly from nowhere. Soon a small inquiring flock has assembled, calling constantly as if to summon yet more birds. As my companion and I stand a foot apart talking, a brilliant male dashes by at our knees. A bird alights on my friend. Everywhere birds are busily foraging on the ground, gleaning food too minute for us to see. They explore the rock fireplace or pass beneath those long flattened logs that form the retaining wall and bench at the front of the lean-to. Quickly becoming acclimated, they enter the lean-to itself to pry around in the dried balsam needles of the built-up bottom. I have watched birds equally at ease in a long, dark, windowless cabin penetrate into its inner-most recesses. Inquisitively, a resplendent male alights on the top of a log, resting at an angle against the rock wall of the fireplace. While the bird watches us preparing breakfast, the lower end of the log, not 3 feet away, burns merrily. We enjoy the birds while we can; indeed next year there will be no enticing crop of cones, and the birds will have vanished. Somewhere, closer to the west, coastwise, they will have located a new food supply.

I want to focus on crossbills because they are here *this* year, and that's a treat. Years go by when the white-winged crossbill (*Loxia leucoptera*) are absent. John James Audubon (quoted in Stone 1937) wrote at Camden (New Jersey) that in the first week of November 1827, "they are so abundant that I am able to shoot,

White-winged crossbill.

every day, great numbers out of the flocks that are continually alighting in a copse of Jersey scrub pine, opposite my window." They were then present in the winter of 1836–1837, and did not reappear until the winter of 1854–1855, when they were reported to be "so tame that they could be killed with stones." Like the great gray owl, the crossbills are northern birds, and northern animals are commonly tame as they have little experience with humans. They come south only sporadically.

I remember seeing the crossbills when I was a boy. Undoubtedly they have been back to Maine since then, but it was not until this winter that I first began to think I might finally have a chance of getting intimate contact by actually finding their nest.

Finding a bird's nest, like making a scientific discovery, often depends upon a good deal of luck. One can increase one's chances considerably by applying standard methods. First, you have got to make sure the bird occurs in your area. You must then sleuth out its specific habitat. You next identify the breeding season, primarily by listening for singing males. If males are singing, then breeding territories are likely being set up and/or mates are being attracted. The often very specific breeding sites then have to be identified within that habitat. After that, you start observing the individual birds, preferably at dawn, when they are most active. Myriad clues give hints of nest-building readiness or progress. These include seeing your bird with nesting material or food for its young in its bill. Finally, you follow your bird, watching its every move, trying to divine its every intention, and all the while you try to be unobtrusive. Finding a bird's nest is a bit like trying to capture a secretive animal, perhaps a water shrew or a pygmy shrew, that may be everywhere yet is nowhere seen. "Finding a bird nest" the nature photographer Eliot Porter (1966) wrote

> is a skill for which all the guide books, all the geographical check lists and life histories, and all the learned volumes on ornithology are of little help. The nest finder must go out into the fields and woods with his wits sharpened to a razor's edge, with all his senses tuned to their highest pitch, and with his mind free from the distractions and preoccupations that burden the society he has temporarily left behind. His consciousness must be focused on the world outside himself, in which he must move without self-awareness. If he succeeds in attaining this rapport with nature, all creatures, as Thoreau said, will rush to make their report to him. He will learn who his companions are, where they are, and what they are about. All their activities will be as shouted declarations, and no secrets will be kept from him.

My expectations are modest. I go with the flow, but the start of my journey, as in any of scientific discovery, starts from precedent. I had seen the birds previously in December and then again in February, and I hoped to see them again now, in March. In December they were still traveling and feeding in flocks, so nesting had not yet started, but by mid-February the raspberry pink males had left their flocks and were singing their loud, musical warbles and trills, while the golden-brown females hopped unobtrusively in the spruce branches nearby. There were mutual chases; the flocks had disbanded and breeding was about to begin. However, I saw no nest-building, and the presence of females out of the nest meant that eggs had not yet been laid. Given that it takes about a week to build a nest, and three or four days to lay three to four eggs, plus about two weeks of incubation, I calculated that now, in mid-March, they should be close to finishing with incubation.

Crossbills raise their young when the seeds of spruce or pine cones are most plentifully available. This often requires them to lay their eggs in the winter. Nests with eggs have been found in New Brunswick, Canada, in the middle of January, and in February near Calais, Maine (Smith 1949). Crossbills are reputed to breed at almost any time of the year depending on any of a variety of different kinds of cone seeds they may find. In contrast, the crossbill's relatives in the family of finches to which they belong, our goldfinches as well as European goldfinches, are the latest-breeding birds; they delay breeding until August, when *their* seeds (thistle seeds) are ripening.

Finches are strikingly colorful birds, but perhaps even more amazing is their unusual and varied bill morphology, adapted for extracting thistle seed, cracking cherry pits, or prying seeds from under stiff cone bracts. Crossbills' bills look misshapen, as from some developmental defect. Their long and slender (for a finch) two-centimeter-long upper bill crosses over a one-half-centimeter-shorter lower bill. By inserting their partially open bill under a cone bract and then closing the bill, the bill-tips separate by about 3 millimeters, applying strong leverage laterally so that the bract is pried

Crossbill prying apart cone bracts to reach seeds.

away from the cone. The seeds under it can then be reached with the tongue. Given their unique bill structure, adapted for extracting the seeds from pine and spruce cones; their wide wanderings over the continent in search of seeding conifers; and their specific timing as to when they nest, crossbills are consummate conifer specialists and they are a truly northern or boreal species.

Most of the nest descriptions date more than a century ago, from the heyday of the great naturalists. These descriptions whetted my appetite. The deep cup nests of crossbills had almost always been found in dense spruce trees from as low as seven feet up to seventy feet high. They are reputedly built of spruce twigs, and variously lined with "wool and moss," "rabbit fur," "moss and animal fur" (Macoun 1909), "felted black wool-like lichen" (Grinnell 1900), and "long black slender tendrils resembling horse hair" so that the nest appears "nearly black." The eggs have been reported as having a ground color of pale blue, bluish-green, greenish-white, or creamy-white and they are marked with scattered spots and blotches of "pale chocolate," "pale lavender," "ashy-lilac," "scrawls of black," and "lines of bay and fawn-brown." Why would nature produce such beauty in such a temporary thing as an eggshell that the bird sits on and hides? The young are "sooty black" and covered with down, and when they gape for food, their mouth linings shine a "scarlet" or "bright purple red." In this case the bright colors are signals. They induce the parents to notice and feed their young, amplifying the begging response.

My anticipated winter excursion to see crossbills, or for whatever looking for crossbills might yield, I had driven to Maine in the night. Parking my pickup down at the bottom of the hill below the cabin, I saw only a white hillock hiding my neighbor's truck; it had snowed much here. I strapped on my snowshoes and was relieved to find the walking easy since the snow was thickly crusted over. The moon was not yet up as I walked by starlight.

A barred owl hooted on York Hill and for a minute or two a second one answered from Larkin Hill on the other side of the valley. It has been a good year for maple, beech, and oak mast; there is a good mouse population. Then I heard only the steady crunching of my snowshoes on the snow. I strained my ears hoping to hear a coyote concert or maybe a saw-whet owl. But the woods soon became eerily silent.

Up at the cabin the snow had slid off the roof on the uphill side, piling up above the back windows and above the top of the back door. Thanks to the downhill grade at the front side of the cabin, I could still get inside. I quickly built a fire and checked the guest log, reading that one visitor had come on February 18 and he had reported "cold, -10°C, and gusty winds. Deep snow. Tough even on snowshoes." A lot of snow had come since he had been here, because I had seen no trace of his tracks while coming up. And then I crashed into bed.

I was almost asleep when I was serenaded wide awake by howling coyotes perhaps a mile off, toward Wilder Hill. After I was again almost asleep, I heard the faint, hollow, almost pulsing booming rhythm of a great horned owl. I sprang out of bed and opened the window to hear it better. The booming came from near the swimming hole in Alder Stream, and memories flooded back of Bubo, my great horned owl, who had followed us down there to bathe in the pools by the rocks. Perhaps it was he. It seems hardly possible, but a sleepy longing is kept alive on occasions such as this, which are not infrequent.

I awoke near daybreak and groggily forced myself up and onto snowshoes, to be out into the woods quickly. I beat the first rays of the sun coming over the ridge by Kinney's Head, then watched them shining golden on the red spruces where I had observed the crossbills singing and cavorting a month earlier. A mourning dove started its mournful, owl-like predawn serenade. Two more called from the west and north. There was not a breath of air, and except for the dove's continuous dawn chorus of coos, it was still.

Minutes later it got noticeably louder. A hairy woodpecker commenced with hollow-sounding drumming. Two others chimed in, hammering also on dead "drums" of wood, sounding a different pitch but the same cadence. Within half an hour of wandering about the woods, I had seen two pairs of red-breasted nuthatches and a small flock of chickadees. I had heard purple finches and pine siskins sing. A small flock of evening grosbeaks flew over giving their clear bell-like calls. A raven perched in the pines and called *rrap, rrap, rrap* ... The rhythm and the cadence sounded like that of my raven friend, Goliath, who has lived with me around the cabin and who has a mate and who nests annually and raises four to five young on a specific branch, on a specific pine tree next to the camp. No crossbills. Not one peep. Could they have left after exhausting the red spruce seed crop? Did the recent big snowstorms destroy their nests?

The spruce trees were still heavily laden with enough cones to make the tops of the trees look brown. However, while spruce cones stay long on the trees, the *seeds* fall out of them as the cones dry and the bracts curl out.

Red squirrel.

A month earlier spruce seeds were strewn about on the fresh snow and chickadees were hopping on the snow foraging for these shed spruce seeds, while red-breasted nuthatches were picking them out of the cones in the tops of the trees. I found one spruce tree with several hundred cones under it on the snow. These cones had been chewed off and dropped by a red squirrel. Warmed by the sun they had then melted into the snow. So this squirrel had not been in any hurry to gather them up after dropping them.

Impulsively I picked up a handful of the cones. All had only the first few bracts chewed off near the base of the cone. The snow surface was littered with cone bracts; the squirrel had been feeding on the cones as it was harvesting them up in the tree. But why did it just drop and leave most of these cones unopened? Back at the cabin later I examined five of these discarded cones. Each had on average 40 bracts, and a full cone has 2 seeds under each bract. Thus, a full cone produces about 80 seeds. The five cones could contain close to 400 seeds, but I found instead only a total of 23, or about 5 seeds per cone remaining. No wonder the squirrels had dropped the cones after sampling them, or not gathered them up later.

How many seeds must a cone contain before a squirrel decides it is not worth the effort to invest more energy and then discards it? A squirrel can eliminate the possibility of sampling the same bract twice, because it must

Hemlock cone.

Closed.

Red spruce cone open and shedding seed.

chew each bract off in order to see what's under it. However, crossbills cannot know for sure which of the 40 bracts on any one cone it has sampled has seeds under it. Since it can't very well memorize which bracts have been sampled and which not, it likely can't help but sample some bracts more than once. It may therefore need to have fairly full cones in order to make the hard work of separating the stiff bracts from the cones worthwhile. On the other hand, prying bracts apart might be easier than the squirrels' way of chewing them off. Had the crossbills left here because the cones had by now shed too many seeds?

This was not a question I could hope to answer, but it did induce me to climb to the top of a spruce tree in order to retrieve a twig laden with cones. The bracts of the cones were opening up, and when I banged limbs that had cones on them I caused showers of seeds to twirl down. I examined ten cones, having 40 to 50 bracts each, finding that seed number per cone (20, 6, 36, 8, 2, 35, 28, 16, 12, 17) was considerably higher than on my previous count of squirrel-discarded cones. Was this an unusual tree, or one that by chance crossbills had not visited? I was obviously not going to answer this question just now, either. But it was a fabulous day with deep blue sky, rising temperature, and the crust was still solid for excellent walking on snowshoes. I'd better make the most of it.

Walking on, I eventually found two other spruce trees where squirrels had been feeding on cones. Ten out of 321 had only a few bracts eaten off at the base (where a squirrel always starts) and the rest of the cones were then left. That is, they were discards.

But seed number per bract this time was high—similar to those on trees. However, most of these seeds had small holes in them as though they were infested by a tiny insect. What was originally a simple question was getting more complicated with every bit of data I collected. It was not just a one-weekend project. My seed counting assured me that spruce seeds were available, though variable and spotty, but I got no insight into the lack of crossbills, and instead raised questions about red squirrel behavior.

Coming off the spruce ridge for a quick lunch at the cabin, I saw a chipmunk scampering off on top of the snow. The chipmunk must have burrowed a long ways through snow to see daylight. It could not hope to find food on the snow, but because it was the mating season for chipmunks, this one came up out of hibernation for another reason.

I rushed back to the cabin to be warmed by the woodstove and to stoke my metabolism with a quick meal of bread and cheese. Rewarmed and refueled, I then headed right back out on snowshoes and walked down the hill toward Alder Stream. Within a few minutes I was on fresh moose tracks. There were many moose tracks, and they were being sprinkled with thousands of snow fleas. Near every moose track the snow was also sprinkled with a strand or two of moose hair, which would likely be used to line bird nests in two months' time. Occasionally, a deer had followed in a moose's tracks, probably because that was easier than plowing their own way through the deep snow. The usual background pattern of red squirrel, hare, and grouse tracks crisscrossing was everywhere. A porcupine left a deep groove in the snow where it had regularly traveled between its rock den shelter and a hemlock tree where it fed at night. Two coyotes had been traveling side by side, a bobcat had traveled alone, and one intriguing track in the spruce woods could have been a pine marten's.

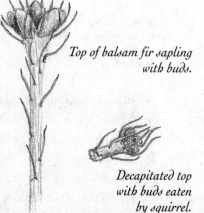

Top of balsam fir sapling with buds.

Decapitated top with buds eaten by squirrel.

I came to a blowdown area of mature balsam fir that had in the last decade become a thicket of young regrowing firs. Strangely, almost all of these young fir trees were missing the top buds of their leading branch, the one vertical twig that would later become the trunk of the tree. What could possibly have happened? I had never seen anything like it. Few tops of young balsam fir trees remained—only 4 out of 56, 3 out of 57, and 3 out of 75—at three different sites. There were red squirrel tracks next to many of the smaller trees where there were fresh chips from chewing and snipped-off tops with their buds removed. Many of the trees whose tops were out of reach of moose and deer were also damaged. So it was squirrels. Normally at this time the red squirrels start to make maple syrup and sugar candy by puncturing sugar maple twigs and coming back to feed after the water from the sap evaporates. But in this patch of these woods there are no sugar maples, so the squirrels have hit on a new snack.

Balsam fir winter buds, harvested routinely by red squirrels.

Female cone buds.

Leaf and twig buds.

Twig with male cone buds.

Tip of a fir twig with all the male flower buds chewed off.

By afternoon the tiny wisps of cloud disappeared, and the sky became a deep cerulean blue. The wind died down and the mourning dove called again. Coming back to the hill I heard a raven in the woods. Its muffed voice disclosed that this bird had food in its bill; it was probably feeding at a winter kill. So I walked in the direction of the call and soon found a raven's track leading to the edge of a stump where the snow was disturbed. I dug there and found buried a fresh piece of red meat that was sprinkled with deer hair. A raven's cache.

Another raven came up to the valley, and as it crossed a canopy opening in the dense spruces, it saw me and shifted its fluid wing strokes to rapid wingbeats that make a loud *swoosh-swoosh-swoosh* sound. The second raven would notice, and did—it heard and flew up in alarm. I now advanced at a trot directly to where the deer carcass would be.

Most of it was under a mound of snow that had been pawed in from about four to five feet in all directions. Huge cat tracks all around. Fresh red meat and chewed-off ribs were visible in a hole dug into the middle of the mound. A mold in the snow next to the carcass showed where the cat, a large bobcat or a lynx, had lain down near the deer, leaving behind a few tufts of its downy soft fur.

I dug the deer out of the snow and found hemorrhaging on the head and in the back of the neck; it had been alive when bitten. The wounds confirmed that a cat had killed this big, perhaps 140-pound, doe with a foot-long fetus inside her.

The raven flew by once more and called. I left then and detoured to check on a porcupine den in a pile of rocks in a dense spruce-fir thicket. Fresh tracks had gone in and out during the previous night. The porcupines were now eating the bark off red spruces and white pines, girdling the trees in huge bare patches of exposed trunk. They had long ago killed off their favorite food, hemlocks. Under one tree they had recently killed I found a regurgitated pellet, along with bird droppings near it. This pellet contained the long straight black guard hair and softer insulating hair of a bear, as well as a few balsam fir needles and a spruce cone bract. It was a raven's pellet. Another winter kill might be nearby and so I immediately searched through the thicket, expecting the source of the pellet's bear hair, but instead of a dead bear I found another dead deer. Both ravens and coyotes had just been feeding on it. No cat tracks here and this carcass was not buried by snow scraped over it. I stopped at the kill site, in my mind's eye seeing the struggling of predators and prey, starvation and struggles in the snow, but hearing only the sweet songs of purple finches, pine siskins, and best of all the elfin song of a golden-crowned kinglet. The kinglet now buoyed my spirits as I enjoyed its sweet refrain and thought of the incredible endurance and luck that brought this bird through the winter, to finally have a hope of seeing spring.

As I made my way back to the cabin, I saw snow fleas collecting ever more by the thousands in the deep holes of the moose's tracks; temperatures were now above freezing. A little spider walked on the snow, and then I found a rust-red furry Arctiid caterpillar humping along on the crust of the two- to three-foot-deep snow as well. Where was it going, and why, with not a green blade of grass in sight? (I took it home and reared it to the adult stage, and identified it as the ruby tiger moth, *Phragmatobia fuliginosa*.)

Finally as I returned to the cabin, there was no more doubt in my mind that most of the crossbills had left. Their nests had probably been destroyed in the recent, unusually heavy snowfalls. I should have known, since crossbills leave even their young in their nests if, in their boom-or-bust cycles, the food supply runs out (Macoun 1909). Only stray wanderers might remain, and as if to be reminded of that, one crossbill came in to camp.

A friend, Glenn Booma, a Winter Ecology student of 1990 who comes often to escape city life in Boston and who has proved his credentials by once trapping a water shrew, came to visit for supper. We lit a fire in the firepit to grill steaks, and as the smoke from the dried maple we had chain-sawed and split the previous fall rose into the early evening sky, a bright strawberry pink (adult) male white-winged crossbill flew in. He fluttered within a foot of my ear and then landed at the edge of the firepit. The crossbill hopped close to the glowing embers and picked at ash. Within another minute it departed as quickly as it had come, leaving us in surprise and wonder.

Scientific discoveries, like most surprises, come by luck, and luck comes by keeping moving and having a keen nose to detect anomalies. One may keep moving, to look for a crossbill's nest, but, idiosyncratically, may find a raven's winter food cache and much more instead. And that is also how weasels find and capture chipmunks in the winter woods, which I will get to shortly.

Here in New England two species of weasels molt from brown in summer into a new white coat in winter in response to day-length changes. They retain, however, the black tip on the end of their tails. (In other warmer parts of their range— weasels are found as far south as Florida—both species remain brown year-round.) Three other northern weasels are also native to New England: the mink, the pine marten, and the fisher. None of the three changes its pelage color. All are members of the Mustelidae, the remarkable family that includes wolverines, otters, skunks, ferrets, and badgers (including the memorably named African honey ratel). Of those occurring locally, all remain active during winter, except the striped skunk, which becomes semidormant and lives off its fat.

Of the two species of weasel that turn white in the winter, the long-tailed weasel (*Mustela frenata*) does not extend far into Canada, whereas the ermine (*Mustela erminea*, also known as the stoat, in England) has a more northerly and circumpolar distribution. In the field it is practically impossible to distinguish these two. Males of both species weigh about twice as much as females, and *M. frenata* are about twice as large as the ermine. Undoubtedly at least some of the "ermine" that used to decorate the coats of European aristocracy was in fact long-tailed weasel. Both came from the trappers in North America.

Weasels, at first glance, seem to be designed wrongly. They are beautifully camouflaged in white, yet that conspicuous black tip of a tail seems an odd, inexplicable anomaly—until experiments showed that hawks easily captured fake weasels that had no black-tipped tail. When the hawks were baited with fakes that had black-tipped tails, however, the birds grew confused, either momentarily hesitating or attacking the tails as though they were the head-end. Other small animals also use such deception-evolved tails. Many lizards, for example, have colorful, conspicuous tails that divert or distract predators. The tail is easily detachable, and starts writhing and flailing after being detached, to divert the predator even more from the rest of the animal that slinks away. Lycaenid butterflies also have similarly distracting and detachable "tails" on their wings that fake out a predator in much the same way a good basketball passer fools his opponent on the court. The butterflies' tails imply to a predator that its prey is about to head in one direction, when it then turns and escapes in the opposite.

Weasels are consummate mouse predators, but they are not restricted to a diet of mice. A 1999 study of least weasels (*Mustela nivalis*), which are native to North America north of New England, and to Alaska, Europe, and Siberia) shows that small rodents (mostly voles) constitute 41 percent of their diet in summer when they also eat birds, eggs, and insects. In winter they subsist primarily on small rodents.

The flexibility of these predators is noteworthy. Weasels readily climb trees; the pine marten in particular is known for its tree-hunting of squirrels. So skilled are weasels that in Wytham Woods, near Oxford, England, the population of tree-hole-nesting great tits was reduced by 50 percent before predator-proof nest boxes were constructed. I've been told by Maine woodsmen that in the winter, weasels will even catch snowshoe hares, a feat I find difficult to imagine as the largest weasel that I have weighed is a *Mustela frenata* male of 283 grams. An adult snowshoe hare weighs five times as much. However, weasels are not intimidated by a size disadvantage. My neighbor near my Maine cabin told me of seeing a white weasel in November in close pursuit of a hare.

The mustelids' ability to kill large prey may involve more than brawn, as evidenced by the finesse fishers display while preying on porcupines, which no dog can subdue or eat. They display a curiosity and willingness to acknowledge the new. I suspect they are fairly intelligent. The long skulls of all mustelids, from weasels to otters, indicate a remarkably large brain volume for such a small animal. According to at least one anecdote that I trust, weasels can count (or at least have a sophisticated concept of quantity) up to six.

For about a year my father had a pet weasel that, when still young, used to ride around in his coat pocket. The weasel came from Bulgaria. Papa had been hiking in the country one spring when he heard a rustle in the dry fallen leaves. He thought he had encountered a brown snake. Instead, it turned out to be a mother weasel followed closely by a train of her seven young, one right behind the other. He picked up all of the cute weaselets and put them into a cloth bag, tied it shut at the top, and put the bag into his knapsack. I presume he'd been trapping small mammals and catching fleas; the cloth bag was standard equipment for retrieving the fleas that jumped off the animals, which he sold to the Rothschilds in London for their famous flea collection.

Papa continued his hike and eventually sat down and opened the knapsack to get his lunch. The mother weasel followed him in pursuit of her stolen offspring. She entered the open knapsack. Papa opened the cloth bag. The mother weasel went into the bag, pulled out one of her young, and ran off with it in her mouth. Even more curious now, he waited. The weasel mother returned, entered the knapsack again, and rescued her baby number two. And so it went, until baby number six. But for number seven she never came back, and it became my father's favorite pet. For a long time after it grew up, it amused and amazed houseguests with the speed with which it could catch a dozen mice simultaneously released into the bedroom. That weasel got even those mice that climbed the curtains. Ultimately the weasel was killed in an accident, the victim of the very features that give these mustelids their edge in hunting success with mice. The weasel, being small and eager to explore hidden spaces, had crawled under a blanket on the bed, and someone had inadvertently crushed it.

Unlike my father, I prefer not to keep weasels free in the house, but opportunities to see them in the woods are rare, and usually fleeting. I recorded a few encounters, including the following. The woods had just been blanketed by a fresh snow on top of a recent crust, and the boughs of the balsam firs were bent low. The oaks and maple twigs formed horizontal lattices on which thick, white pillows of snow had accumulated in hours of windless silence. On this Christmas Day 1995, the forest floor was still clean of tracks and the immaculate snow surface sparkled with pinpoints of sunlight glinting off mirrors made of millions of individual snow crystals.

Weasel in its winter coat.

A movement caught my eye as I scanned the scene. It was a weasel. Against this glistening background, the white weasel, with its black-tipped tail, looked almost lemon-yellow, especially toward its hindquarters. I froze as the animal came closer, dragging something. Within ten or fifteen feet of me, the weasel dropped its prey, raised itself tall on hind legs, and eyeballed me for a few seconds. Apparently satisfied that I wouldn't interfere with its operations, it then dropped down, grabbed the still-limp chipmunk in its mouth, and continued on its way over a rise among the trees.

The local chipmunks had been denned for two months already. Like other ground squirrels, they would be immobile and in torpor, at least some of the time. In such a state they would be unable to escape any predator that could reach them.

The eastern chipmunk builds a twelve-foot burrow system that includes a nest chamber some three feet underground, food-storage chambers, escape tunnels, and one main work hole. The weasel, however, has a long skinny body and very short legs that allow it, like a dachshund (a German name that translated means badger-hound) bred to capture badgers (dachs) in their dens, to enter rodent tunnel systems, such as those of chipmunks.

Chipmunks lay up winter food stores, allowing them to avoid or reduce the amount of time they spend in torpor, which is when they are most vulnerable to predators. Every fall the chipmunks near our house spend days on end running to the bird feeder, filling their cheek pouches to bulging capacity with sunflower seeds, running into their burrow and unloading, and returning for more. In years of plenty of oak, beechnut, and sugar maple mast, the chipmunks also haul in the seeds of those trees. The more food stored, the more time the chipmunk will remain warm-bodied, awake, and vigilant in winter. Indeed, chipmunks are not nearly as prone to spend the entire winter in torpor as other ground squirrels, which don't lay up a larder. In 2000, a heavy mast year in northern Vermont, I saw the local chipmunks frequently emerge above the snow, making many trips to the bird feeder throughout the winter. The next year, when there was almost no beech or oak mast in northern Vermont (but a large crop in southern Vermont), they were already absent above ground in late September. It seems that like many a millionaire, a chipmunk in winter facing unlimited amounts of food can never quite get enough, yet when they have little they know how to get by.

I waited about ten minutes for the weasel to get out of view and settled with its prey before I started following its tracks. Why had it bothered to drag its heavy load? Why didn't it feed on the chipmunk right where it caught it, presumably inside its snug warm den?

Weasels live throughout the Northern Hemisphere and even up into the Arctic. They are active all winter. They must contend with intense cold, yet they are small, skinny, and poorly insulated, all of which facilitate rapid rates of heat loss. To compensate, their resting metabolism is twice that of other animals their size. Yet they have small stomachs and unlike their cousins, the striped skunk, they put on little body fat. As a result, they have to eat more food per day than any other winter-adapted animals.

Yet for all their seeming design flaws for retaining heat, they are in fact superbly designed rodent predators. Weasels need to be small and skinny to enter the chipmunk's tunnel, and balance their energy budget by behavior. Radiotracking studies show that most of their time in a typical twenty-four hours in winter is spent eating and resting. Weasels need no permanent den, nor do they need a large stomach, because after reaching the rodent's nest they use their victim's nest for their own and curl up into a ball to conserve energy supplies while feeding about five to ten times per day. Finally after finishing their meal and again in need of energy supplies, they sally forth on their next hunt.

After dragging its prey about thirty yards in a fairly straight line, the weasel I was watching had climbed with its freshly killed chipmunk up onto a small knoll. There the tracks suddenly circled and zigzagged back and forth in a small clearing. Clearly, the predator had been searching for something in that several-square-yard area; the tracks still showed the drag marks of the dead chipmunk. As I said, the crust was thick, and the fresh snow on top was soft, maybe an inch or two deep. I suspected that the weasel was searching for a very specific hole, possibly the entrance to a den now covered with crust.

I noticed the tracks leading off from the knoll to the base of a nearby small hop hornbeam tree. Along the trunk of that tree, where the ice crust was thinner, the weasel had finally descended into the snow. I next waited for about fifteen minutes, hoping for it to come back out. First I sat quietly; then I squeaked like a mouse in distress. Still no weasel. I then dug down through the nearly two feet of snow to the unfrozen leaf-covered ground by the hornbeam tree, continuing to dig and following a tunnel in the snow. The snow tunnel (much like tunnels I later learned were routinely made by chipmunks) led to the trampled area where the weasel had left its many circling tracks on top of the crust, and that was where I found the entrance hole of a tunnel leading into the ground itself.

I put some loose snow over this entrance and left. When I returned later in the day the hole was indeed opened and the fresh weasel tracks, without accompanying drag marks, led away into the woods. The well-fed carnivore had left.

I checked once again the next afternoon following another morning snowfall when the old tracks and the holes were obliterated. There were still no new tracks. Nor did new weasel tracks appear later.

This was apparently not the weasel's den. It was probably the usurped den of a previous chipmunk victim that this weasel remembered. The most recent victim was likely caught *outside* its den; why else would the weasel have dragged it so far through the snow? That is, this chipmunk had probably not been in torpor and it got caught anyhow, because if it had been torpid it would have been inside its snug nest, which the weasel would have used while consuming its meal. Thus, staying warm to remain alert does not necessarily guarantee survival for the individual, at least not for this chipmunk. An individual's odds are determined by its own specific circumstances, and small idiosyncrasies in the life circumstances of different species almost guarantee different strategies as well. Each animal's existence is balanced on its own often conflicting mix of contingencies.

A large part of adapting to the winter world
involves creating a suitable microclimate. Birds
wear insulating feathers, mammals fur, and we
wear adaptive clothing. But a number of animals,
primarily beavers, bears, humans, birds, and some
insects take a step beyond insulation, building
nests or dens that supplement or take the place of
body insulation.

In all animals, den construction is constrained
by building materials, energy requirements for
heating and cooling, defense, and accessibility. The
Anasazi Indians of the southwestern United States
built their homes high on inaccessible and defensi-
ble cliffs, choosing locations where an overhanging
ledge offered shade at noontime in summer and
exposure to direct sunshine in the winter when the
sun is low. To the north and in Europe and Asia,
when wood was not available, Ice Age hunters built

huts framed with mammoth tusks and covered them with skins and sod; the early Eskimos did similarly, substituting whale bones for mammoth tusks. They of course invented the igloo, that marvel of simplicity and efficiency. Using hard-packed, fine-grained snow cut with a knife into blocks roughly twice as long as high, a man could build a house in less than an hour by spiraling the blocks upward, each slanted slightly inward.

In our own expansion into the winter world we at first relied on the retention of body heat, which is still the main source of heat in an igloo. Then we used fire as a supplement. The first use of fire for heating our hearth is attributed to European sites of about 500,000 to 100,000 years ago, when Ice Age hunters leaned poles against the cave entrance and covered them with hides. Not much changed in the way of heating (except for containing the fire in a fireplace) until Benjamin Franklin invented the Franklin stove, followed eventually by the invention of central heating. Instead of building nests, Ice Age people stayed warm and alive, much like a chickadee or a kinglet does in winter—by metabolically burning fat from the animals they killed.

Among possibly the first nests on earth, and still among the most impressively engineered ones, are those of insects who have been perfecting their building techniques for maybe 300 million years. Every autumn after the leaves fall in New England, I see the nests of one species of wasp, the white-faced hornet (*Dolichovespula maculata*), hanging conspicuously in the trees. Each nest is started in May by a single female that has hibernated through the winter. She uses her mandibles to scrape fiber from the dead wood of a branch, mixes it with her saliva for a papier-mâché slurry, and then makes tissue-thin strips of paper by adding one load at a time on the bottom edge of a slowly growing globe that will house her and eventually her entire brood of hundreds of eggs, larvae, pupae, and adult offspring. As her daughters come of age, they help their mother construct the nest, laying on layers of white, brown, and gray papers, differently hued depending on the wood. Despite its thinness, the paper does not dissolve in the rain. It remains durable all summer long and through the winter. As my experiments suggest, a good soaking seldom wets anything beyond the first layer of the paper nest.

White-faced hornet nest with two wasps adding paper to the bottom edge of the outermost sheet.

As the wasp family grows and more room is needed, the insects enlarge their nest by recycling the paper walls from the inside to make new, larger ones on the outside. A nest starts out in May no bigger than a walnut with just one paper shell, and ends up basketball-size by late summer having about a dozen layers of paper insulation surrounding almost as many horizontal combs with pupae and larvae, hanging one above the other inside. Each air layer between successive sheets of paper acts as insulation, and the temperatures inside the nest stay between 29° and 31°C even in cold weather down to 5°C, as the wasps shiver with their muscles powering their wings to maintain their own body temperature above 40°C. The heat loss from the body then heats the nest and young.

We are recognizably humans for only about 3 million years, but definitive birds have been flying well over 100 million years. At first their nests were, like those of their reptilian ancestors, simply depressions scraped in the ground or hiding places for their eggs and/or young. As birds became ever more metabolically active, in part to become creatures capable of rapid and sustained flight, they at the same time became warm-blooded and the ability to produce heat ended up being not just an option but a necessity both for them and their young. Expanding northward or into seasonal environments then generated selective pressure for their nests not only to be receptacles for their eggs and young, but also warm shelters.

The nests of northern and late-winter nesting birds, such as gray jays, raven, crossbills, and golden-crowned kinglets, are superbly insulated structures that are essential for these birds' lives in their respective environments; the eggs and young must be kept warm, while minimizing heating costs.

The golden-crowned kinglet's nests are built by suspending strands of spiderwebs, bark strips, and caterpillar silk into a hammocklike configuration within the hanging twigs of a spruce or fir. Moss, lichens, and strips of paper birch bark are then incorporated into the bottom and wall of the nest. The resulting cup-shaped structure is suspended into twigs by its lips and its walls. It is insulated with snowshoe hare down and with small bird feathers, numbering 2,486, 2,674, and 2,672 feathers in three nests where individual feather counts were made (Thaler 1990, pp. 83–84). The nests are hidden from view from above, and partially protected from rain and sun by overhanging twigs of spruce or fir. However, regardless of how elaborate, almost all bird nests are built for only one-time use, and a good time to find many of these discarded nests is in the winter.

Bird nests are records of genetically encoded behaviors. Each bird performs unlearned nest-construction behavior, duplicating fairly precisely what its parents had done. Some nest-construction behaviors are relatively simple. Geese, for example, pick an appropriate nest location, and then simply lay their eggs there. Often the female pulls in nearby vegetation from all around and banks it against her body. She then sheds her belly feathers and tucks them under her for a nest lining. In the Canada geese I've watched in a bog by my house, this down-lining of the nest occurs only after or when the eggs are laid. (Reduction of the goose mother's belly insulation could probably be hazardous when she swims in ice-cold water, but it may be necessary on the nest, so that body heat becomes available to the eggs she sits on.)

Most birds collect specific nest material from a wide area, and they execute specific behaviors both to find and work with it. Bicknell's thrushes in the mountains of Maine and Vermont line their nests with the rhizomes of horsehair fungus. One year the pair of tree swallows in one of my birdboxes in Maine built its nest with grass from beneath the birdbox then lined the nest exclusively with white duck feathers. The nearest white (domestic) duck was at least three miles distant.

There is some flexibility in nest construction (but not much). A tree swallow at my house in Vermont built a nest almost *exclusively* of feathers (87 large white ones presumably from chickens, and one small black one from a raven, to be exact). House wrens used the same nest box immediately after the swallows left the eggs that had rotted (after they stopped incubating during a week of cold, early spring weather). But the wrens built a nest of 675 twigs that filled the box nearly to the top. This nest had no feather lining, although for their second successful brood that they raised in the same box that season, the wrens replaced the nest that I had removed with one made of only 40 twigs and lined with 185 pieces of grass and rootlets and 52 feathers of many kinds of birds.

Male ravens build their nest by holding large thick sticks they break from trees with their bills. With rapid head-bill shakes, they vibrate the sticks against or on the substrate of ledge or tree branches at the chosen nest site, until they anchor there. After the nest platform is built, the bird habitually perches in the center, where fewer sticks then become anchored, and a nest cup with a high rim develops. Once the basket of sticks is completed, both members of the pair then upholster it with bark, moss, grass, and animal fur. Almost all nests in New England contain deer hair, sometimes a dense felt of it. Snowshoe hare, bear, and moose hair are also commonly used to cushion the eggs and help insulate them. The female alone incubates for twenty-one days even at the typically subzero temperatures in February and March.

Robins build a symmetrical hard mud cup that is surrounded by loose debris and lined with soft grass. The whole process involves instinctive responses. Two nests I watched being made were each almost completed in two days. On the first dawn, the female robin deposited a haphazard pile of grass, twigs, and birch bark strips at the future nest site. At dawn the next day, she started bringing mud, making one trip after another, at one- to three-minute intervals for five and a half hours. During the first hour or so the male followed the female but did no work; then he perched in the nearby woods and sang quietly, while she worked. No work was done in the afternoon. By the third morning the female made only a few trips, and she was unaccompanied by her mate. Without exception after depositing the mud, she squatted down and vibrated her body for a second or two; then she got up, turned a few degrees to right or left, and repeated the squat-vibration. She performed as many as sixteen of these routines after dumping a single billful of mud or other debris. Her nest-shaping did not seem to be attributable to a conscious plan. On one long beam under a shed, I found where a robin had built a two-cup nest, apparently because she got started turning at two similarly appearing spots on the initial pile of debris that had spread too far laterally on the smooth beam. I found a phoebe who had built a similarly misconstructed nest also on a long even beam where it might have been difficult to fix the precise nest location. In real estate, location is everything, especially in birds.

Although among geese and robins the female alone builds the nest, in some species (principally wrens and weaverbirds) the nest is built primarily by the male. Such male-made nests initially serve as sexual attractants. The female chooses the nest and through it, the builder, and she signifies her choice by adding the nest lining according to her own finicky specifications. I recall watching African spotted weaver-bird males in a colony, all dangling from their half-finished nests and beating their wings and calling to show off. When prospecting females entered a nest to inspect, the male owner went wild in frenzy. In the northern New England woods, perhaps one of the best examples of this is the winter wren, which builds its nest hidden deep under the roots of trees that have been blown over by the wind. The wren's nest is a snug little cavity with walls camouflaged with a lattice of moss and conifer twiglets. It has a quarter-inch entrance hole at the front. A male wren may build several of these nests in an area (I found a sedge wren who built six in his territory), but only the one chosen by the female will be lined with fur and feathers. The "false nests" apparently not only give the female more choice but may also prevent unmated males from moving into the neighborhood.

Weaverbirds of many species literally weave hanging fiber bags that resemble our textiles. However, one species, the sociable weaver (*Ploceus cucullatus*) of Africa (a close relative of our local weaverbird, the common suburban English sparrow) builds communal nests that in size and shape superficially resemble human-built thatch-roofed huts. Generation after generation of birds add to these apart-ment houses until they may weigh several tons and fill out the entire top of a large tree. The apartment complex shields the individual nests within from the sun and from extreme temperature fluctua-tions in the bird's desert environment of southern Africa where nights are commonly cold and days are commonly hot.

Aside from protection from the elements, an important ultimate or evolutionary reason for such astounding diversity of nests and nest locations is predation. By far the greatest vulnerability a bird experi-ences is often in the long egg and nestling stages. Nest predators include other birds, as well as snakes and mammals. Nest inaccessibility works as an adequate defense for some, but most species rely on hiding their nests to reduce predation. However, for effective hiding, much depends on what *others* are doing.

In the winter woods, much is buried and hidden, but more is revealed. Curtains of snowflakes drifted over my face, and the snow already piled into cushions muffled my footsteps. Two feet of fresh powder had already fallen on solidly frozen ground layered with fallen leaves. Recently those leaves had obscured the views through an otherwise opaque forest of green. Now, at the end of January, I could see *through* the forest from one large maple tree to another, to beech trees, root tip-ups, and beyond to an open bog bordered with viburnum thickets. I was looking for bird nests that had in the summer been almost impossible to see because they were then enveloped in leaves. They would now show up as dark silhouettes with white caps. (The ground nests obviously will be missed, but most are almost invisible even without a covering of snow.)

We often find nests by sheer luck. But like nest predators, we improve our chances if we know where and what to look for. That is, present success depends on previous success. Similarly, predators also develop search images based on their previous experience. But by focusing on the familiar nests they may miss those that are different or at other than the usual locations. Variety of nests and nest sites is often a key to nest survival in many species. A simple thought experiment shows why. Suppose every species of small songbird were to tuck its nest behind the loose dead flaking bark of dead balsam fir trees, as brown creepers do. If every bird species were to choose such a site, then it would not be a hiding place. Instead, these sites would soon *indicate* where to find eggs and tender nestlings, in the same way that the Golden Arches now show us where to find a hamburger. What applies to the nest site applies to the nest itself as well.

Programming for precise nest construction yields species-specific nests, and the more *different* or exotic the nest appearances are for different species, the less any one would stand out to predators. Like a nest predator, during my winter walks I also hunt using search images for the nests with which I'm familiar. One of the first of these nests I looked for on this winter day was that of the red-eyed vireo. Each vireo nest is suspended on its upper rim in a horizontal fork of a thin branch, usually in the understory of a hardwood forest. Each nest looks like a light-colored round lump, a little like a hornet nest, but when examined from up close you see how the nest is suspended at its lips by what appear to be spiderwebs and strips of birch bark.

On this day, when I found and looked at the nest of the red-eyed vireo closely, I was surprised to discover pieces of gray paper from the nest of a white-faced hornet that had been incorporated into the outside nest wall, along with the always present birch bark and spiderwebs. The white-faced hornets' communal nests, unlike birds' nests, are added to and grow throughout the whole summer. By fall, however, when the worker wasps and drones die and the mated queens hibernate underground, the wasps' nests are abandoned. Pieces of the nest can then be removed with impunity, but merely touching an occupied nest, as is well known and as I confirmed in experiments where I agitated these wasps, instantly results in an exit of hornet guards that attack all moving objects nearby and sting them mercilessly. I presumed therefore that the piece of hornet-nest paper that was incorporated in what seemed like an outside decoration must have originated from an old, abandoned hornet nest, since a hornet would not hesitate to attack and sting, and likely kill, a bird or other small animal that disturbed the communal nest.

I seldom find a hornet nest. The vireos must have traveled far to get their wasp-nest paper. Was this a typical red-eyed vireo nest? I remember the first red-eyed vireo nest I ever found. I was eleven years old, it was my first spring in Maine, and I had climbed the big sugar maple tree next to the road on the Adamses' farm, where I spent an ecstatic summer. The tree was one that Floyd, the father of the family, had tapped in the spring for sugar sap. I do not, of course, recall if it had wasp-nest paper on the outside, but I wish I did. But being alerted by this nest, and the memory of it, I then made a special effort to examine all red-eyed vireo nests that I found in Vermont and Maine. All twelve that I found had at least some white-faced hornet paper on the outside. Since a vireo nest is globular and resembles a white-faced hornet nest in shape, it seems plausible that incorporating wasp-nest paper would enhance that effect and keep the premier woodland nest predators (squirrels, chipmunks, and blue jays) from coming close. Indeed, a slight jiggling of a branch with a wasp nest causes the inhabitants to attack en masse—a response no squirrel or jay would want to risk. Even if one of these common bird feeder chow-hogs is only slightly deterred by something as seemingly trivial as a nest decoration, that might still make a big difference in the long run. I estimate that chipmunks and red squirrels alone raid well over 60 percent of *all* songbird nests in my woods.

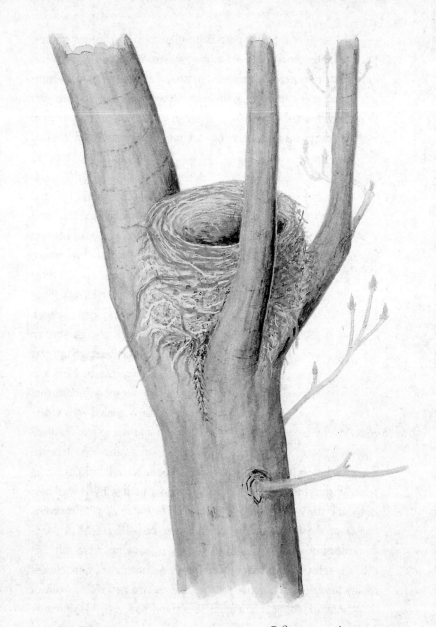

Redstart nest in young sugar maple at the edge of a swamp.

The linings of the vireo nests were all of thin, firm fibrous strands, but the kinds chosen differed among the many nests. Long reddish dead white pine needles were used in two nests. Hairy-cap moss fruiting stalks were employed in another. Other materials included very thin grass stems, the dried stems of sugar maple flowers, fine strips of ash bark, the rachis from decaying fern fronds, and sedgelike fibers that I was unable to identify.

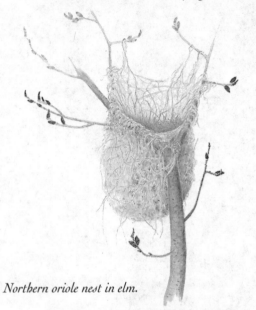

Northern oriole nest in elm.

As I left the hardwoods where I found the red-eyed vireo nest, I descended onto a beaver bog where the frozen pond surface was bordered with already long-yellowed dry sedges. Patches of light brown cattails were giving way to spiraea bushes, and arrowwood thickets that then graded into alders and young maples. Redstarts and least flycatchers at the latter site wedged their nests into the vertical forks of young trees. Within the cattails I found a half-dozen red-winged blackbird nests and three common grackle nests. They were all within several inches of the ice in the wilted, now-dry cattail leaves. The long-discarded bulky nests of catbirds and cedar waxwings were still visible in tangles of thin twigs where in the summer they were almost totally hidden in the dense foliage. The catbird nests were built of long twigs and robust plant stems, but the almost invisible signature that distinguishes them from similar nests (blue jay, rose-breasted grosbeak) is strips of grape bark. The one cedar waxwing nest I found was, as is typical, decorated with green moss. The low twiggy spiraea bushes revealed the flimsy grass nests of chestnut-sided warblers and also a yellow warbler nest made of cattail seedfluff felted together into a deep fluffy cottony cup. Goldfinches had left their plant-down-upholstered nests in the more open-standing young trees and bushes.

Catbird nest in willow with nightshade vine.

An elm tree along the beaver bog edge held the almost perfectly preserved baglike nest of a northern oriole. Its weave of tough plant fibers that surely rival those of hemp (which I found in decaying milk-weed stems) allowed it to be suspended in the wind-whipped tip of a long thin twig where nest-raiding squirrels seldom travel. Near ground level I also noticed a jagged hole through the thin, tough bark of a rotten gray birch stub. It was the entrance to the nest cavity excavated by a black-capped chickadee. All Paridae, or tits, the family to which chickadees belong, hide their eggs in enclosed spaces. Chickadees have small weak bills yet they commonly hammer out their own nest hole in soft rotting wood. They sometimes use preexisting cavities instead, such as those made by a woodpecker. Within them they build a soft nest of moss, hair, and feathers.

Building material often dictates architecture and gives new options for unique nest sites. Barn and cliff swallows, for example, can place their nest on inaccessible cliffs (usually in dense communities of dozens of nests), essentially constructing their own often reusable stucco birdhouses out of mud mortar that remind me of Anasazi cliff dwellings, making one wonder if Anasazi were inspired by cliff swallows. Inside the swallows' reusable mini-birdhouses (which are generally considered the nests) they build their nests of grass and feathers. Because of the mortar exterior that can be glued onto any solid substrate, many swallows can nest as readily on a cliff as on a barn wall. In contrast, bank swallows dig holes into sandbanks and within them build their grass-and-feather nests at the end of sandy tunnels. Tree swallows use instead the old nest holes woodpeckers have hammered into trees. Chimney swifts make a shallow nest cup by using their saliva to glue one little twig after another in parallel, forming a thin bare shelf onto the chimney wall. Their relatives, the cave swiftlets in Asia, have dispensed with twigs and build their whole nest out of saliva. (The coagulated hardened bird spit is considered a dining delicacy, since it is an expensive Asian restaurant item.)

Chestnut-sided warbler nest in spiraea, roofed over with Clematis *seeds. With seventy-six cherry pits stored inside. A deer mouse larder.*

Chestnut-sided warbler nest in spiraea bushes.

As I continued my walk and looked around for fascinating nests in the woods and bog on this snowy day, I found nothing quite as exotic as swift or swallow nests. Instead I found several nests belonging to chestnut-sided warblers. As is typical for this species, the nests were small, round, cup-shaped structures built of fine grass and situated in low spiraea bushes. To my great surprise, however, I found one nest that was domed over and it had a small round entrance hole to the side, reminding me of a wren's nest. But this nest was just another typical grass nest of a chestnut-sided warbler, with a superstructure of plant down that had been added later by some other animal. Inside were 76 black cherry pits. After this surprising find I looked more closely and found another similarly revamped domed-over nest. This one, however, belonged to an alder fly-catcher, and it contained 483 milkweed seeds, each with its fluffy parasol completely removed. I also found a goldfinch nest filled to the brim with about 1,200 uniden-tified small black seeds, 254 milk-weed seeds, and 1 sunflower seed. (I planted some of the small black seeds and they grew into common roadside ragweed, whose pollen is a common allergen.)

Goldfinch nest in red maple sapling.

The three bird nests had, I suspect, been taken over by deer mice. Unlike other mouse and squirrel nests that serve as nurseries for young in the spring and as shelters from the winter elements for adults, these partially arboreal mice had cleverly rebuilt nests to serve as grain bins for winter food larders. Bird nests are often passed on among other bird species. For example, wood ducks, buffleheads, and mergansers depend on the old nest holes made by pileated and formerly ivory-billed woodpeckers. By providing safe nesting sites, woodpeckers are thus keystone organisms for a vast assemblage of birds the world over, including many owls, parrots, parids, flycatch-ers. But the recycling of the bird nests for food storage by the mice was, as far as I know, a previously undescribed behavior.

Cedar waxwing nest in arrowwood.

The mice made me wonder why almost all the other bird nests I found were destined to be abandoned. Why don't birds reuse their nests for winter shelter? Even the nonmigrants that stay all winter such as chickadees, goldfinches, purple finches, and blue jays don't reuse their nests or build new ones for shelter or for food storage. Yet most birds are skilled and eager builders. They construct amazingly elegant and functional nests with which to attract mates and in which to incubate their eggs and rear their young, but would they build even a crude shelter from the killing cold? Nothing doing! Well, *almost* nothing doing. There are, as almost always in the winter world, blatant exceptions.

The verdin (*Auriparus flaviceps*) is one. This mouse-gray bird which is halfway in size between a kinglet and a chickadee is a year-round resident of the southwestern United States and northern Mexico, where temperatures in winter can drop easily to 0°C at dawn. Not only is the verdin an apparent exception to most birds, it is a conspicuous exception because it builds *numerous* bedroom nests at *any* time of year.

The verdin's breeding and roosting (or bedroom) nests are both globular structures of about 20 centimeters in diameter. They are shaped like a small squirrel nest, but have only one entrance. Breeding nests take about six days to construct, while roosting nests are finished in just four days. Verdin nests are elaborate structures of three layers. The outermost layer is made of tightly interlaced thorny twigs that solidly anchor the nest into the tree; typically a mesquite, smoke tree, paloverde, or catclaw acacia. The middle layer is a compact mass of matted leaves, feathers, lichen, moss, and other various materials. Finally, the nest interior, especially in winter nests, is lined with soft fluffy material such as feathers from various birds, plant down, and spiderwebs. The winter roosting nests are larger and better insulated than summer nests.

Verdin nests are not only used by verdins. Glenn E. Walsberg, at Arizona State University, discovered a verdin nest in a shrubby desert wash that, during December 1989, was being used as a communal roosting place for fifteen to sixteen black-tailed gnatcatchers (*Polioptila melanura*). For five nights Walsberg observed these birds as they dispersed during the day, arriving back at the nest in the evening. Walsberg again counted them as they left the nest the next dawn, but this time the nest remained unoccupied for two nights. Obviously, the birds had alternate roosting sites and were traveling en masse from site to site.

Walsberg recorded internal nest temperatures near 30°C at night, even as air temperatures outside the nest dipped to near 0°C at dawn. Overnighting in the communally elevated nest temperature of 30°C would have cost each bird about 2.5 times less energy per night to keep warm. However, total energy savings were even greater because the birds were not only taking advantage of increased air temperature surrounding themselves, but also heating by direct body contact. I could not in the Maine woods, expect to discover anything resembling Walsberg's amazing findings of gnatcatchers using a verdin's nest, but I wondered why not. Surely a kinglet *could* use similar shelter.

By late afternoon the snow was coming down so thick that it almost blinded me. I passed the old hemlock tree with a honeybee nest that I'd found by lining bees the previous fall. A hollow ash tree where I had once seen a raccoon hole up in January stood near, but hearing the loud *kek-kek-kek* of a pileated woodpecker, I veered toward a huge dead elm instead. These woodpeckers had for several years in succession carved out their nest cavities in this tree. I saw three cavities, one above the other and spaced about four to five feet apart. Woodpeckers go to great trouble and effort to always make a new nest cavity each spring. There is then a yearly progression of fresh empty apartments for flying squirrels, crested flycatchers, barred owls, and potentially bats.

When I reached the woodpecker-made nest holes, I again heard the pileated's *kek-kek-kek-kek*, reverberating through the woods, and then I saw the big gorgeous black bird. White wing-bars flashing, it sliced through the storm and landed on the far side of a sugar maple near me. It jerked its white-striped head with flaming red crest toward me from behind the shielding tree. I backed off, and the bird then flew to the elm and entered one of the old nest holes, possibly one of its own used for nesting in a previous spring.

Woodpecker holes are used by other birds for shelter in the winter as well. I have often found red-breasted nuthatches overnighting during the winter in old downy or hairy woodpecker holes, and the pygmy nuthatch (*Sitta pygmaea*), which often travels in small flocks, overnights during the winter in tree holes with its many companions for added warmth (Knorr 1957; Guntert, Hay, and Balda 1988), as do Carolina chickadees (Pitts 1976) and occasionally also bluebirds (Frazier and Nolan 1959).

It is one thing to find existing shelter and make use of it; it is quite another to expend time and energy making a shelter for future use. Surprisingly, some other northern woodpeckers do just that. In late November and early December when temperatures are dropping rapidly and the first snowstorms blanket the woods, I often hear steady tapping unlike a woodpecker's more intermittent rapping for food excavation. Following the sounds to a decaying tree or thick tree limb, I find the ground and/or snow below littered with small light-colored wood chips. The head of a downy or a hairy woodpecker invariably appears at a round hole, then shakes to release a billful of shavings, and quickly tucks back in to resume hammering. At first I thought these birds were confused, perhaps by misreading the pho-toperiod—the hours of daylight versus dark that many animals use to keep track of the seasons—to begin nesting a half year early. How-ever, the finished holes excavated in the late fall or winter (in more decayed wood than nest holes) were invariably used by the birds that made them for overnighting sites; I flushed out the bird in the evening by tapping the tree, but it quickly reentered the hole to spend the night there. (Woodpeckers use no nesting material beyond a layer of dry wood chips, but the nest cavities provide obvious protection from the cold.)

Since it may take a pileated woodpecker or a sapsucker up to two weeks to hammer out a cavity in a tree whose interior is softened by fungus, one wonders why other winter birds don't build "regular" nests for shelter since such nests can be built in several days. For example, a pair of golden-crowned kinglets constructs its well-insulated nest of moss held together by spiderwebs in just four to six days (Ingold and Galati 1997). Might golden-crowned kinglets make winter nests as a solution to overnighting at subzero temperatures; or could they use nests made by other species?

I've never seen a kinglet carrying moss or any other nesting mate-rial in winter, and I think it is unlikely that these birds build winter nests for shelter. But if they don't build winter nests, then *why* not? Do they not have the time because they are living too close to the edge? I have never seen a kinglet take even a one-second break from winter foraging. Maybe when their priority is to survive the day, then making an investment that doesn't pay off until a later date makes no sense. Additionally, if the bird has to keep moving fast and continuously all day in an environment where food is patchy, then at the end of the day (when it may have traveled far), it may not have either the time or energy to relocate its nest. Having found an area of good food, it may be the better bet for a kinglet to stay put and forage until the last moment of daylight, and in fact that is what kinglets do. Finally, a winter nest would have to be designed differently than the open cup nests for rearing young. Although modifications of nest structure are possible, a new nest design might be too large an evolutionary hurdle for them to surmount.

The verdin's and the woodpecker's winter shelter nests are only slight modifications of their breeding nests. Their nests evolved first as receptacles for rearing young, and that same design then fortuitously made them preadapted for use as winter shelter later on. The nest of a finch or kinglet, in contrast, is cup-shaped. It is not preadapted for shelter in winter. Shelter nests require a cover. Cup nests in winter would quickly fill with snow.

There is no law that says a priori that a bird can't switch from making one type of nest in one season to a different type of nest in another season. It is just one of those things that is unlikely, because the more specialized the programming is in one direction the more it precludes a new specialization in another direction.

Some effective shelters that birds make in the winter are much easier to construct than nests. The grouse that dives down into the deep fluffy insulating snow to escape the cold also modifies the snow by creating a cave at the end of its tunnel where it comes to rest. This shelter is for one-time use only. The bird may spend a night or a day or more in its shelter (depending on the weather), but it never returns to it. The next night it makes another shelter. However, if snow is absent, the grouse roosts in the cover of a dense coniferous tree instead.

As I was searching in the bog for the discarded summer nests of birds, I almost took for granted the most obvious winter shelter of all, the lodges of the resident muskrats and beavers. Beavers (Chapter 12) build lodges at the edge or directly in the pond, by piling up logs and sticks, mud, and sod into a steeply conical heap of five to seven feet above water level, and then excavating from underneath. A suitable lodge can be built in the fall, but old lodges are also refurbished and may be used by generations of beavers. I measured a lodge built the previous fall, and it reached seven feet above water (or ice) level and it was fifty-two feet in circumference. On this particular lodge almost all the sticks were light yellow, because the beavers had only recently eaten off the bark. Mud had been heaped onto the sides (but not the tip top) of the mound compacting and cementing the lodge together. In the winter when the mud freezes solid, the communal beaver lodge becomes practically impregnable to wolves and other potential predators.

The entrance to a beaver's lodge is through a water lock at the bottom so that the beavers have a safe den site even in the summer. This was not lost on John Colter, the famous Yellowstone explorer who in 1809 escaped a band of pursuing Blackfoot warriors by diving into a beaver pond and hiding in the animals' lodge. One parched summer a year or two ago, when the local beaver ponds had dried up and the water seal to a beaver's den was gone, I also entered one of their dwellings. I carried a flashlight, to have a look around. The den I examined contained two platforms. One was slightly above the other, and in both the floors were liberally strewn with small debarked twigs, the remnants of take-home meals. (There were no remnants of fecal matter.) The chambers were roomy enough for me to be able to turn around, but it must have been a tight squeeze for a mated pair of beavers, together weighing about one hundred pounds. Add their two to five yearling offspring, and an equal number of young from the current year, and it must have been a tight, cozy fit. The space seemed awfully small for a crowd of perhaps a dozen beavers to spend half a year there in total darkness. Beavers do not hibernate and must therefore be quite tolerant of one another, as they would almost literally be rubbing against each other all winter. By spring, however, the yearlings get the wanderlust—or the cold shoulder as the parents evict them. After the ice melts off their ponds, I find these yearlings as common roadkill on highways that cross brooks or streams.

Even larger animals may build shelters. Usually before snow starts to cover the ground, a bear (or a family of a sow and her previous yearling cubs) chooses a den site for hibernation. Grizzlies dig tunnels into a hillside and excavate a cavity at the end. Black bears excavate depressions under a pile of brush, or the roots of a fallen tree, or use a rock cave or a hollow tree, and sometimes make or use no cave at all. The typical den cavity, if there is one, is about five feet wide and two or three feet high and upholstered with leaves, grass, and other debris raked in from the surroundings. Cedar bark and conifer twigs may also be bitten and torn from nearby trees and carried in. Yearling cubs help their mother build the den that they share with their mother. A bear's den retards convective cooling by the wind, but temperatures inside the den are not much different from air temperature outside; bears rely mainly on their fur for insulation, which doubles its insulative capacity in winter. Once settled in, they don't feed for seven months, living off their body fat until late spring. In January, sows give birth to two to three naked cubs, which don't hibernate. They snuggle up to their mother and suckle for the three additional months that she hibernates all the while losing about twice as much body weight as males. Both sexes, however, stand a 99 percent chance of surviving the winter. The den stays clean because the sow does not urinate or defecate all winter. The cubs do, but despite a greatly depressed appetite the mother eats their feces.

The biologist "Bearman" Lynn Rogers, studying black bears in northwestern Minnesota in the 1970s, routinely visited them in their dens. "Most of those I visited in dens were wakeful enough that they lifted their heads and looked at me," he wrote. "Although, in general, they seemed less sensitive to danger than they had been in summer, some were moderately aggressive. . . . Some did not wake even after gentle prodding and jostling." And: "In one case, on March 27, 1970, I accidentally fell on a six-year-old female in her den. She didn't wake up for at least eight minutes, even though her cub bawled loudly and I began prodding her. On January 8, 1972, I tried to hear the heartbeat of a soundly sleeping five-year-old female by pressing my ear against her chest. I could hear nothing. After about two minutes, though, I suddenly heard a strong, rapid heartbeat. The bear was waking up. Within a few seconds she lifted her head as I tried to squeeze backward through the den entrance. Outside, I could still hear the heartbeat, which I timed (after checking to make sure it wasn't my own) at approximately 175 beats per minute." A bear's heart rate during the day is normally 50 to 90 beats per minute, although the heart rate of a sleeping bear in winter may decline to as low as 8 beats per minute. I wonder who was the more surprised by the visit: the researcher or the bear.

Other than biologists, few humans have entered bears' dens to find out how cozy they might be. But in one case I am intimately acquainted with, a rabbit-hunting beagle wandered under a brush pile in the Maine woods, which happened to be the den of a black bear with her two cubs. The beagle attempted to retreat, but every time the dog tried to crawl out, the bear dragged it back inside. The sow acted as though the beagle was one of her cubs. The owner finally retrieved his dog unharmed, but only after the bear was tranquilized with a dart gun and the dog had then been denned for two days longer than it intended.

The second inadvertent entry into a bear's den that I heard about involved a man. And it ended with a hastier exit. The man apparently broke through the snow and fell into a cavern on Ellesmere Island that happened to be the den of a polar bear with cubs. The heavily padded man (lucky for him, he was wearing his microclimate!) was quickly and summarily heaved back out with one swat of the bear's great paw.

In the woods on the hills near my cabin in Maine, I commonly see what are popularly called bear "nests" up in beech trees. These I have entered, or examined closely, without being molested. They are platforms of branches that the bears pull toward themselves from all around to strip off the beechnuts when the trees still have their leaves. Leaf-shedding by trees is an active process in preparation for winter (to reduce ice- and snow-loading), and a broken and killed branch does not shed its leaves because the tree's physiology is disrupted. As a consequence, where bears have been foraging for still unripe beechnuts in late summer one sees in November what superficially look like giant squirrel dreys (a name for squirrel nests) with many dead leaves. Structurally, the bear dreys are almost identical to chimp sleeping nests, the best constructions that any of our closest living relatives are able to produce.

One April I found young northern flying squirrels (*Glaucomys sabrinus*) with still-closed eyes, and I adopted one of the litter. Fed on Similac baby formula with an eyedropper, the tiny waif grew quickly. It often slept in my shirt pocket when I carried it to my office and occasionally to the campus dairy bar, where I enticed it out onto the countertop to lap up ice cream. Outgrowing my pocket, it later lived in a spare bedroom, where it slept all day in a hollow log. When I entered the room after dark, it ran up to a ceiling beam and jumped off to glide through the air and land on my chest with a light thump. When jumping from hundred-foot trees, northern flying squirrels can glide over three hundred feet, given suitable slope and wind.

Like my father's pet weasel, my squirrel died in an unfortunate accident. I had put sprigs of geranium into a jar with water. One evening when the squirrel was free in the living room, it crawled down the cut geranium stems and drank from the water they were in. A little later the little animal was retching: it had been poisoned by the plant's chemical defenses. The next morning my charming pet was stone dead. Animals that to us appear to have astounding toughness, such as surviving in the winter world, are also extraordinarily fragile, each in its own way.

Northern flying squirrels are common across North America from the Canadian maritimes all the way to Alaska, and they survive the harshest winters. Whatever it is that flying squirrels do to live through northern winters it does not involve the usual tricks of storing food, getting fat, or hibernating. Furthermore, my tame ice-cream-lapping flying squirrel notwithstanding, these animals are normally strictly nocturnal. One might predict instead that they "should" try to avoid night activity to avoid low temperatures by then resting in their snug nests, yet in the wild they sleep away the day even when temperatures are reasonable. They come out of their snug nests only when the sun goes down and temperatures dip sharply. I have no answer to why they do this, but a comparative perspective gives hints of selective pressures that have made flying squirrels nocturnal.

It is likely indicative that *all* of the about thirty species of the world's flying squirrels are nocturnal, while *none* of the one hundred or so species of day-active squirrels are adapted for gliding flight. The fact that none of the day-active mammals are fliers or gliders cannot be attributed to dietary specializations. Does it concern predation? Gliding flight saves much energy for moving around, yet it makes the animal conspicuous to predators, with the additional drawback that the wing membranes compromise agility. (Bats, the best mammalian fliers, are among the poorest runners.) Maybe squirrels flew to save energy, then had to become nocturnal to escape predators, then had to fly even more, because the noise of scampering on the forest floor would be the owls' hunting cue. Hence, the need for energy economy in these mammals would positively reinforce a nocturnal lifestyle that encourages gliding and flying.

Flying squirrels don't leave being active at night up to mere chance; their circadian clock ensures that they get up and going only *after* sundown. That does not mean that they do not heed light cues from the environment. They do use light cues to synchronize their internal clock to keep to the daily twenty-four-hour rhythm so that they can get up and go out of their dark daylight hiding places soon after it gets dark outside. How then do we know whether a night-active animal becomes active after dark because it is the right *time*, or merely because it is *dark* then (and vice versa in a day-active animal)?

Flying squirrels were important in answering this fundamental question. They are one of the first mammals that were shown to be able to become active at the right *time* independent of external cues. The pioneering and now-classic experiments that revealed the fascinating world of chronobiology in the southern flying squirrels (*Glaucomys volans*), and subsequently in almost all other organisms examined, were conducted by Patricia J. DeCoursey from the Zoology Department at the University of Wisconsin.

DeCoursey's research was based on sixty-eight squirrels that she trapped and raised in Wisconsin. The squirrels were individually housed in cages, each equipped with a running wheel mounted on a bicycle axle. An eccentric cam attached to the axle momentarily closed a microswitch circuit at one point in each wheel revolution to leave a mark on a chart moving at a uniform rate of 18 inches per day. A continuous record of both the number and time of wheel revolutions were thus displayed for later analysis. The continuous record over many yards of paper was cut into daily strips that were aligned by time and then pasted one day beneath the other in sequences of days extending over months. From these records DeCoursey could determine within two minutes when the squirrel had run in the twenty-four-hour cycle and how the activity on one day compared with others.

From the stack of numerous twenty-four-hour records one on top of the other, she saw at a glance that, not surprisingly, the flying squirrels are night-active (unless, of course, they can safely come out to the dairy bar for ice cream). They began running shortly after dark, and then they ran either sporadically or almost continuously (depending on the individual) until dawn, when they ceased all running activity until the next night, or they have two activity periods, one right after dark, and another before daylight.

The above now and perhaps even then rather prosaic results were the preconditions for the real experiment, when DeCoursey next put the caged squirrels into *constant* darkness. Would they now run continuously or sporadically? The answer was: neither. Surprisingly to DeCoursey, each squirrel ran on the wheel at nearly the *same* times as it had previously when it had experienced a twenty-four-hour light-dark cycle. That is, the squirrel knew when it was time to be active because it apparently consulted an internal timer. Skeptics cautioned that perhaps the squirrels were instead responding to some unknown external or exogenous signal that was associated with evening, rather than keeping to their previous schedule by using an internal or endogenous time sense.

Ultimately, DeCoursey proved with her squirrels that the timing originated *internally*, and it was no small irony that her best proof came from the squirrels' small errors in timing. For example, Squirrel Number 131 on average started to run (in total darkness) every 23 hours and 58 minutes, plus or minus 4 minutes, while another under the same dark conditions in the same room ran 21 minutes later each day; i.e., it had an activity cycle of 24 hours and 21 minutes. That is, under constant dark conditions, one squirrel lost 2 minutes each day while another gained 21 minutes per day. Within ten days of "free running" in constant darkness, one squirrel started activity 20 minutes before external evening, while the other was then 210 minutes late, or 3.5 hours out of synchrony with the external world. If both squirrels had consulted an exogenous or external timer, then they both would have run as if responding to the same drummer; they would have kept the same time.

The squirrel's clock-running speed is genetically determined, but the time at which the beginning of the animal's running activity is *read off* is determined by frequently resetting the clock in reference to an external signal. In squirrels, the signal to which their internal clock is synchronized is the moment of lights-out. We now know, for day-active animals, that light hitting the eye causes the pineal gland of the brain to reduce its production of melatonin, a sleep-inducing hormone that is normally produced rhythmically, on an approximate (but not exact) twenty-four-hour schedule. Hence melatonin pills to combat jet lag. A flying squirrel would have to take them in the morning. When DeCoursey reintroduced a one day light-dark cycle in their environment, then the off-schedule squirrels that had been "free-running" in continuous darkness reset their activity regimen to again start running right after lights-out the next day. Ordinarily the squirrels therefore reset their clocks when subjected to the normally occurring light-dark cycle. Superficially they act as though they respond only directly to darkness or light, and without the experiments that is all one could know. DeCoursey could, of course, have waited to make her observations at total eclipses of the sun. But it would have delayed her conclusions because of difficulties for replicating of observations. Experiments involve making things happen and then applying keen observations of the results.

DeCoursey's demonstration of internal time-keeping was simple, elegant, and irrefutable. It brought a closure to the debate of whether or not a mammal had an internal circadian clock, and it opened up an area of research in cellular mechanisms. We now have volumes of information on circadian clocks since DeCoursey's experiments in the 1950s, and the information is becoming of tremendous relevance to medicine. For example, effective dosages of many drugs depend strongly on the time in our own circadian rhythms when they are administered. The molecular mechanisms whereby circadian clocks operate have lately been traced to a number of genes, and the most popular "model organisms" in which they are now studied are no longer flying squirrels but mice and fruit flies.

The circadian clock has many potential uses. It allows hibernating ground squirrels, for example, to measure the daily light-dark durations, and from that data the squirrel can derive information about the changing seasons. Correct seasonal responses are crucial for winter survival. Indeed, the circadian clock mechanisms are necessary for all organisms that must prepare for winter, whether by pupating (insects), migrating (insects, birds, some mammals), or hibernating and physiologically preparing (most northern organisms).

BEING ABLE TO GLIDE from tree to tree is a very efficient way of locomotion, but in flying squirrels the shift to nocturnal activity is costly from the perspective of energy supplies required to keep warm. Energy is saved by gliding, but the ability to glide precludes the laying up of fat stores, such as practiced by their relatives, woodchuck and other ground squirrels, that may get to be obese by fall. Also unlike ground squirrels, northern flying squirrels do not avail themselves of the huge potential energy savings of torpor, since they also don't have a buffer of energy stores in body fat or food caches, nor change into a more insulating winter coat. As regards energy balance in winter, much seems to be stacked against them. One wonders what solution they might have that counterbalances their numerous presumed shortcomings for winter survival.

I am on the lookout for squirrel nests in my search for overnighting sites of kinglets, so I habitually bang on any tree that holds a nest to see if a kinglet seeking shelter might fly out. All the northern flying squirrel nests I've found were in dense spruce-fir thickets. I've never chased out a kinglet, but on occasion I've been rewarded with seeing one or two individual flying squirrels pop out of a nest, glide off the nest tree, and land on a neighboring tree. Assuming that the squirrels spend half or more of their time in winter in their nests, nest insulation should be of great relevance to energy balance. One nest that I examined in December 2000 was an unfinished framework of dry spruce twigs that contained no lining. Confined between several upward-bending branches, it had probably been abandoned before being finished because the space was too small. It showed, however, that the squirrel starts its nest structure by first making a globe of dry twigs, then inserts the lining. That December I found six other nests that had the same magpielike frames of small dry twigs but that did however contain the nest proper. (One had been torn open, and nest lining had been pulled out.)

The nest linings varied from nest to nest. In one I found a mixture of moss, lichens, grass, and shredded birch bark. In two the lining was almost exclusively finely shredded birch bark. In a fourth it was almost all moss. In a fifth it was exclusively shredded cedar bark, and

Flying squirrel nest, covered by cushions of snow.

FLYING SQUIRRELS IN A HUDDLE

in the sixth the lining was in two distinct layers of shredded birch and cedar bark. (Many cedar trees in these woods show evidence of some of the outer bark having been stripped off, presumably collected by squirrels although bears also collect cedar bark.) When thoroughly dried this last football-size nest weighed 17 ounces, 12 ounces of which was lining, with a thick 8-ounce shell of densely packed usnea ("old man's beard") lichen and a 4-ounce layer of soft, finely shredded cedar bark within that. A good choice— the Northwest Indians used such shredded cedar bark to diaper their babies.

Even after heavy rainstorms the insides of the nests remained dry. Normally in winter these nests are also insulated on top when they are roofed-over with cushions of snow. All the nests had two entrances, one each on opposite sides. These entrances were not visible. They were, like the elastic ends of our mittens and socks, closed. Thus, in structure, each nest was like an old-fashioned hand muff. (In none of these, nor in seven additional red squirrel nests, was there one speck of bird feces, making it unlikely that they serve as kinglet overnighting sites.)

To get a rough idea of whether the flying squirrel's nest indeed affords much insulation, I heated a potato to simulate the body of a squirrel and examined its cooling rates. At an air temperature of -13°C, a hot potato (60°C) cooled to only 42°C in thirty-five minutes when within the nest, and to 15°C in the same time period when outside it. My rough experiment only says that the nest indeed affords effective insulation. Of course the value of insulation would be much greater in wind, and it would be even more effective in a snow-covered nest. Furthermore, a squirrel, with its downy fur and a bushy tail wrapped around itself, would lose heat much more slowly than a potato. And the slower it cools, the less energy it would have to use up to shiver and maintain a stable and elevated body temperature.

In Jack London's story "To Build a Fire," the newcomer to the

North was ultimately killed because he got his feet wet. He broke through thin ice under a thick insulating layer of snow on Henderson Creek. His fire that was snuffed by the avalanche of snow under a spruce only made it impossible for him to correct his initial bad luck, or mistake. Ironically, in an insulated sock, mitten, or a squirrel's nest, a tiny bit of moisture is far more dangerous than deep cold; because wetness destroys insulation. Thus rain, at near 0°C, can be lethal, while snow at -30°C can ensure comfort because it won't wet and destroy insulation. Without dryness, all lifesaving insulation is for naught, and nest construction or placement must provide for it. Nowhere was this more evident to me than when examining a gray squirrel's nest in winter.

Gray squirrels' nests, or dreys as they are often called, appear as haphazard brush-piles of leaves and twigs when we see them piled up high up in trees. All fall and winter I saw one in the branches of an oak tree along our driveway. In mid-January after a heavy rainstorm, the nest blew down, and when I examined it I found it to be anything but haphazard in construction. It was a functionally crafted thing. The outside layer of the 30-centimeter diameter globular nest was of red oak twigs with leaves still attached. The twigs had therefore been chewed off the tree during the summer. Inside this rough exterior I found layer upon layer (twenty-six in one spot where I counted) of single flattened dried green oak leaves. The multiple sheets of leaves served as watertight interlocking shingles, because the nest was dry inside. The leaf layers sheltered a 4-centimeter-thick layer of finely shredded inner bark from dead poplar and ash trees. This soft upholstering enclosed a round, cozy 9-centimeter-wide central cavity. I could not imagine a more efficient functional design from simple common materials. However, not all gray squirrels' nests are as natty as this. Many that I have inspected *were* mere piles of junk, as though they might have been fake nests to distract predators so that the *real* nest could escape being raided.

Gray squirrel nests incorporate leaves, unlike hawks' nests.

Nests require effort to build, and not all squirrels bother to build one, as I found out with a little help from four of my friends. In the winter of 2000 we saw fresh signs of red squirrels almost everywhere we looked in the spruce forests of Maine where these squirrels live. Yet, we found few red squirrel nests. I wondered if (as is reported in the literature) their winter nests are underground, since I found many red squirrel tracks leading into underground tunnels. I saw numerous tunnels leading under roots of a big rotten pine stump and thought that if a red squirrel nest is anywhere, it should be here. Would this nest be less insulated than that of a flying or a gray squirrel?

Inasmuch as biology is a sterile undertaking until one gets hands-on experience, five of us armed ourselves with spades, pickaxes, plain axes, saws, and a digital thermometer, and then after a good breakfast approached the stump that was in the spruce thicket opposite my swimming hole in Alder Stream. The three entrances going into the ground under the stump had been used within the last day. Bracts of red spruce cones lay in piles on the top of the stump and had been recently chewed. All signs were promising.

It was 8 A.M. and a brisk -14°C on the morning of December 21, 2000, when we started excavating. After only five minutes of work, following tunnels in the spongy duff and soft rotten wood, we were apparently getting somewhere because a red squirrel shot out from one of the three exit holes. We dug deeper and also farther around the periphery of the stump, pulling off huge chunks of frozen humus that, like a carapace, covered the almost dry duff and soil underneath. Then we dug all the way down onto the bedrock under the stump. After about another half hour (and the first signs of skepticism from my friends), a second red squirrel darted out. Now freshly motivated, we dug ever harder, and after another hour and a half we had thoroughly excavated a fifteen-square-foot area all around and under the stump.

We found no food stores and no sign of a nest. Negative data is not considered good evidence, and is generally not reported. However, we had dug all this up so thoroughly that the negative data sure felt like a positive result: No nest. Given the lack of tracks around the stump in the morning, the two squirrels had spent the night in the tunnels under this stump whose spongy humus felt warm to the touch (mostly because it was dry, but measured -0.02°C). The several dry maple leaves that we found in our excavation may have been carried down by the rodents in a weak motivation to build a nest, but in their heavy winter coats they would probably not have needed one.

Flying squirrels may also not bother to build much of a nest when they can snuggle up next to other warm bodies. On November 19, 2000, I was banging yet another tree, this one a dead red maple in the woods near my home, in my continuing quest for birds overnighting in tree holes. I happened to look up in time to see a flying squirrel scamper to the top of the twenty-foot stump and stop there, as if frozen in place. Its flat tail was flush against the bark,

*Flying squirrel in the hole
made by a hairy woodpecker
in a dead sugar maple.*

and it didn't move a muscle. I saw then a second squirrel peeking out of the tree hole above me. I looked again, and saw two *more* scampering up the tree, one behind the other. As my companion and I walked around the tree, the top squirrel jumped off, gliding about fifty feet to the bottom of a live red maple. Seconds later, another squirrel "flew" off in another direction. We left quickly, because we did not want to disturb them further. It had snowed a couple of days earlier, and there were deep nightly frosts. It stayed below freezing all day. Would any of the four return to their shelter?

The next day around 3:00 P.M., after it had snowed hard, I went back to the tree, hoping to see them leave their communal den. I hunkered under the low spreading branches of a red spruce, and there I waited until 4:45 P.M. (forty minutes past sunset), when it was too dark for me to see any more. According to DeCoursey's experiments, the squirrels should have known it was time to get up, even in the constant darkness of their den, yet I saw no squirrels exit the hole.

Four days later I went to the stump again. This time I lightly jiggled it as I had done before. Nothing came out of the hole. Then I banged the tree hard with an ax. One came out. No more. I banged it again, harder. Three more came out. So—they *do* come back to the same place. However, a month later, on December 17, when I checked again, no amount of banging raised a squirrel. I climbed up and examined the hole. Surprise! There was no nest at all, and the hole was only three or four inches deep. It contained dry rotted wood and several tiny wisps of dried green moss that must have been carried in. The four squirrels would have nearly filled the whole cavity, and either there was no need for nest insulation or there was no room for it.

From tracks, I knew that these or other flying squirrels were still nearby. My cabin at the edge of a one-acre clearing was within about three hundred feet of where I had seen the four. One seldom sees the tracks of a flying squirrel in the woods, when they land on tree trunks rather than on the snow. But on March 16 that winter I saw where a flying squirrel didn't quite make it across my acre-sized clearing the night before. The squirrel had hopped out into the field from the south, climbed the maple in the middle, and then twenty snowshoe lengths (sixty-five feet) farther it hit the field again, almost at the edge on the other side. Another, similar flying squirrel track commenced across the clearing from the east, also in the direction of the same big birdbox that I had put up years earlier for some kestrels. The two tracks converging at the sugar maple tree with the birdbox were a clue I could not ignore.

I hit the tree with an ax. One flying squirrel with huge black eyes and soft gray pelage popped its head out of the birdbox. After I started to climb the tree I saw *three* heads looking out. No—it was four! Coming near the nest box itself, I saw several more squirrels climb out and scamper up ahead of me. Definitely more than five. I counted again—four, five, six—and when I was up under to the box itself I saw even more climb up to the very tip top of the maple. They lined up one behind the other, like a queue of planes taxiing down the runway waiting for takeoff. A couple were still close enough for me to touch. One jumped off, flying toward the field, then veering in midair, changing direction while still airborne and gliding to the right. It made a perfect landing at the base of another maple at the edge of the field. I counted again—there were nine more squirrels on the tree with me. Ten squirrels in all! I reached into the birdbox and *felt* a flimsy structure of shredded plant material that was warm to the touch. No more squirrels. Not wanting to disturb the animals on the tree just feet above my head, which had pressed themselves immobile onto the tree trunk, I quickly climbed down and then watched as one after the other of the nine squirrels ran headfirst back down the maple trunk to rush back into their birdbox. The tenth that had jumped stayed away for the time being. All the squirrels were adult, and it was close to mating time. I had even seen enlarged testicles on one who'd been inches from my face.

I learned later that the winter sleeping aggregations of flying squirrels had been described, although the northern flying squirrel had not been reported to bunch up to the same extent as the southern species. Curiously, the communal aggregations are sex-specific (Osgood 1935; Maser, Anderson, and Bull 1981). The squirrels huddle for warmth, but why should males not huddle with females, or vice versa?

When I revisited the birdbox in early May right after the snow had melted in the woods, it was empty, at least of squirrels. When I reached in to pull the flimsy nest structure out to examine it more closely, my hand encountered a deep, foul layer of slimy material that was all to easy to identify. Apparently the squirrels had used the nest box not only as a sleeping place. There was not a bit of lichen in the nest, although lichens are a significant portion of the winter diet of flying squirrels, and lichens were a main component of some of their tree nests that I had found. Thus, although *some* flying squirrels in winter in effect live in a gingerbread house where they may find insulation, others opt for body-warmers and/or a convenient indoor toilet.

HIBERNATING SQUIRRELS
(HEATING UP TO DREAM)

On several winter mornings in 2002 at dawn, when I sat at my desk looking out into the woods, I saw three gray squirrels emerge, one after the other, from the same leafy nest high in a pine tree. Within a few minutes the trio left unhurriedly, traveling on the bare winter branches of the maples. Like tightrope walkers they balanced on the slim branches and then acrobatically jumped from one tree to the next. They fed on the buds of broad-leafed trees after snipping off the terminal twigs (which they dropped), on sunflower seeds when they were available in the bird feeder, and on acorns on the trees or on the ground. In the spring, a red squirrel has a nest with young in one of my birdhouses. And these are not my only squirrel neighbors.

Within less than a hundred yards from my home in Vermont, and within a mile of my cabin in the Maine woods, live two more species of squirrels, in addition to the flying, gray, and red squirrels. All descended from a common ancestor more than 60 million years ago. They diverged and specialized on different kinds of food. They are what they eat and they hibernate depending on what they eat.

The most common, conspicuous, and noisy of the local squirrels is the little red *Tamiasciurus hudsonicus*, also called pine squirrel in parts of its range. It is the "sentinel of the taiga" as William O. Pruitt Jr. calls it in a little book titled *Animals of the North* that I have long treasured. It leaves signs of its presence everywhere: cone bracts of pine and spruce freshly strewn over the surface of the snow, cone cores discarded on a log where tunnels enter the base of an old pine stump. Almost every fresh snowfall is quickly followed by a new sign, and the perpetrator of that sign will likely be perched on a branch next to a trunk above your head. The cheeky little *chicoree* (still another name for *T. hudsonicus*) will let loose with a loud sputtering chatter or a *churrrrrr* that resounds throughout the forest. This will usually be sequenced to a long series of harsh staccato chatter, accompanied by flicks of its fluffy tail over its head and thumping with its hind feet for emphasis. Red squirrels are emphatically active at any month of winter. They appear not to hibernate at all. However, during periods of extreme cold, the woods are silent, and they hole up for days at a time in their subterranean burrows under a stump or tree roots.

Once underground they are almost fully protected from the cold. At least some of their populations, particularly out West, make large caches of seed cones, and with that food they can presumably continue to be active. Yet, cone crops are not reliable every year, and in Maine there are many years, such as the winter of 2001–2002, when I've found no caches at all. At these times they feed on the buds of spruce and fir (see Chapter 3). Although there is no guarantee that they do not become inactive with lowered body temperature for a few days, if need be, their emphasis is on fighting the cold by storing food if they can, finding alternate food if they have to, seeking shelter, and growing a thick, rich-rust-colored, insulating fur coat in winter.

The red squirrels' temporary retreats into tunnels and dens in the winter at subzero temperatures contrasts with the behavior of the other four local squirrel species. Of these, the larger-bodied gray squirrel (*Sciurus carolinensis*) and the much smaller northern flying squirrel (*Glaucomys sabrinus*) stay above ground the whole time. Two others, the eastern chipmunk (*Tamias striatus*) and its very much larger cousin the woodchuck or groundhog (*Marmota monax*) absent themselves from the cold snowy world above ground for weeks, months, and even to half the year. In general, most ground squirrels hibernate all or most of the winter, whereas tree squirrels, which can still find food on trees, don't. The stark differences in overwintering biology within this one group of related animals shows that hibernation is less a strategy of avoiding the cold than of what they eat, of weathering famine.

Hibernating chipmunk.

The local squirrel showing the least tendency to hibernate is the now-often suburban gray squirrel. It is active through all months of winter. In the absence of the largess of sunflower seeds at bird feeders, these squirrels will dig through shallow snow to recover acorns, nuts, and maple seeds stored in the fall. If seed crops fail them, they then feed on tree buds and sometimes bark. Food storage, snug leafy well-insulated nests, and large body size give them enough energy resources and means to conserve body heat so that they do not need to hibernate.

Chipmunks are "true hibernators." Like other ground squirrels, they spend most or all of the winter in a subterranean nest where they curl up, cool down, and become torpid. However, they are not torpid all winter long. If they were, then they would not need to lay up food stores to fuel body heating. Torpid animals don't eat. The chipmunk's large cheek pouches indicate an ancient evolutionary commitment to storing food. I do not know how many seeds a chipmunk usually packs into each of its two pouches—I easily inserted sixty black sunflower seeds through the mouth into just one pouch of a roadkill. Chipmunks seldom fail to fill both pouches on any visit to my bird feeder, whereas the gray and red squirrels never carry away even a single seed. Anything they eat, they have to eat in place.

By late fall after a good sugar maple seed year, or after finding a well-stocked bird feeder, chipmunks take trip after trip fully loaded, and all trips lead into the hibernation burrow system that has special granary chambers. These food stores are especially needed in March when the snow is generally still deep. It is then the mating season and the male chipmunks burrow to the surface. There is no new food yet above-snow, but traveling on the crust is easy and those little ground squirrels with the most stockpiled food from the fall can then be the most single-minded in their pursuits.

Normally I don't see a single chipmunk all winter long. They stay underground, entering into periods of torpor. But torpor is an option, not a necessity or a rule, as was apparent in the winter of 2000–2001 when throughout Maine and Vermont we had an exceptionally large mast crop in the fall. The sugar maples, red oaks, and beech all simultaneously produced bumper seed crops, whereas in many years they produced no seed at all. It was also a winter of exceptionally deep snow. Yet, despite the frequent storms that winter, the chipmunks came to our feeder all winter long.

A chipmunk's availability of stored food affects whether it remains fully active or enters full torpor (Panuska 1959). But entry into torpor also requires a cold stimulus. Chipmunks are light sleepers; handled torpid chipmunks invariably become roused (Newman 1967). When that happens, their metabolic rates increase as much as fiftyfold within an hour. In contrast to the torpor induced by starvation, as in a nonhibernator close to death, the hibernating chipmunks' low body temperature is not passive. At an air temperature of 0°C they *regulate* their body temperature near 6°C, rather than at 37°C when they are active. At air temperatures above 15°C, however, body temperature of hibernating chipmunks is no longer regulated, passively increasing with increasing air temperature (Newman 1967).

Like chipmunks, northern flying squirrels as already mentioned also do not fatten up for winter, nor do they put on a thick insulating fur as red squirrels do. Nor do they lay up stores of food. Instead, they solve the energy problem by huddling in groups in snug nests. Even at -5°C outside the nest, the temperature within the nest is not yet low enough for them to have to shiver to keep warm.

Unlike the eastern chipmunk, some ground squirrels from the western American mountains and deserts enter into hibernation torpor not in response to cold. They begin to hibernate in the hottest, driest part of the year and then continue to stay torpid through the winter (Cade 1963). These squirrels enter hibernation regardless of temperature, and also regardless of the absence or presence of food and water. One of these species, the golden-mantled ground squirrels (*Citellus lateralis*) has gained fame for the revelations that have emerged for the timing of its hibernation, through the experimental work of Kenneth C. Fisher and his student Eric T. Pengelley. Their subject animal, unlike eastern chipmunks, does not store food but instead fattens up prior to hibernation. So much to eat, so little time. How do they know when to begin? They consult an internal calendar.

Calendar-type timing was suspected when Fisher and Pengelley noted that their squirrels kept under constant light and temperature conditions in the lab at the University of Toronto stopped eating and drinking and went into hibernation in October, at the same time that those outside exposed to the natural environment did. For four consecutive years in one experiment, the squirrels' cycles of feeding, fattening, and hibernation torpor all continued in the absence of all external cues, in somewhat shortened annual cycle of 324 to 329 days rather than 365 days. The cycling of behavior and physiology occurred in squirrels held at year-round constant temperatures of both 0°C as well as at 22°C in the lab. At a constant 30°C they no longer hibernated, although they still maintained their annual cycle of feeding and fattening that is normally associated with hibernation. These dramatic results showed that the squirrels consult an internal timer, and in analogy with the previously known circadian or daily rhythms, Pangelley and Fisher coined the word "circannual" (*circa* = approximately, *annum* = year). Such circannual calendars have since been demonstrated in the timing of bird migration and in other subterranean rodent hibernators. They play a role not only in preparation for deep hibernation but also in arousal from it. The groundhog or woodchuck is an especially well known example.

The groundhog is a large ground squirrel that according to the legend popularized by or of the old Pennsylvania Dutch settlers has an amazingly precise internal calendar. At least one of them, Punxsutawney Phil, emerges punctually from his burrow every year on the second day of February (in Pennsylvania) (at the precise time the news media start to roll TV cameras) to check if he can see his shadow to decide whether or not to go down and sleep for another two weeks.

A groundhog's survival does indeed depend on accurate scheduling: The squirrel must synchronize its life with the availability of veggies. If possible, it feeds on lettuce, carrots, peas, beans, and other freshly picked produce. Its natural food, grass and weeds, will do only if it can't get into a garden. In either case, food is available for only about a third of the year. Furthermore, greens don't store well in the constant moisture under snow and in underground burrows. They'd quickly become a moldy mess. (A relative of rabbits, the pica that lives in mountain areas where there is more wind and dryness has hit on a solution. It gathers greens and dries them to make hay, which is stored and later eaten throughout the winter.) The groundhog has a different solution. It converts the summer greens not to hay, but thanks to a prodigious and adequately preprogrammed appetite, to thick rolls of body fat.

Obesity has its advantages, such as when the animal can be safely inactive in its den. For the rest of the time obesity makes the animal a considerably more attractive meal to predators, all the while compromising its speed and agility. To minimize its *duration* of obesity, the groundhog must maximize the *speed* and extent of becoming obese. To be successful in this endeavor, it delays fattening until near the end of the summer. So, it must not only know what to eat, it must also consult a calendar as to when to start eating as if life depended on it. Thus, as in the golden-mantled ground squirrel, a circannual clock is vital for its winter survival.

FOR PROBABLY THE MOST remarkable story of hibernation and winter survival of any mammal, I turn now to the arctic ground squirrel (*Spermophilus parryii*). This tawny-gold and gray ground squirrel with small white spots is larger than a chipmunk and smaller than a woodchuck. It is the northernmost mammalian hibernator across the Canadian and Siberian tundra. For eight months of the year this squirrel curls up into a ball close to the ice of the permafrost, and maintains a body temperature at or below the freezing point of water. Brian M. Barnes and colleagues at the University of Alaska at Fairbanks have for many years tried to decipher how these animals survive. They have studied them in the field at the Toolik Lake station at the foothills of the Brooks Range, and in enclosures and in the lab at Fairbanks.

Like other ground squirrels, this species digs hibernation burrows and builds underground nests. However, because of the permafrost of their environment, the squirrels cannot dig deep enough to escape the subzero temperatures of winter. Instead, in late summer and autumn, when the temperatures are merely freezing, they dig down into the soil. The temperature surrounding them declines through fall and winter and continuous records of body temperature in squirrels equipped with radio transmitters indicate that body temperature of torpid squirrels declines in parallel with the sinking soil temperature. Remarkably, however, as burrow temperatures continue to decline to -15°C by December, the squirrel's body temperature does not continue to decline, to -15°C. Instead, it stabilizes at between -2° and 2.9°C. That is, the torpid squirrel's body temperature is then no longer allowed to be passive. It is regulated some eight to nine degrees lower than a hibernating chipmunk's but twelve to thirteen degrees above soil temperature. No other animal had previously been shown to regulate its body temperature near 0°C, much less two or more degrees below the freezing point of water. Furthermore, the squirrels did not turn into blocks of solid ice as their core body temperatures declined to as much as -2.9°C. Barnes wondered if they might have antifreeze. To find out, he removed blood plasma from hibernating squirrels and tested its freezing point in the lab. The blood turned to ice at approximately -0.6°C. It therefore did *not* contain antifreeze. These results deepen the mystery of winter survival: Why should the blood freeze in the lab but not in the animal? The riddle is not yet solved, but the best tentative explanation so far is that the squirrels supercool.

Pure water has a freezing point of 0°C (32°F). Adding one mole (molecular weight) of a substance to a liter of water lowers its freezing point by -1.86°C. Although pure water and solutions of specific concentration have precisely predictable freezing (and melting) points, it is sometimes possible to lower temperatures *below* the *predicted* freezing point without having ice crystals form. Such solutions are said to be "supercooled." Supercooling occurs due to the absence of "nucleation sites"—places where ice crystals can begin to grow. The best nucleation sites are other ice crystals. Thus, if one adds an ice crystal, say a snowflake, to a vial of pure liquid water that is supercooled to -10°C, then the whole vial full of water will turn instantly into a solid block of ice. But this ice won't melt until it is heated to 0°C. This difference between freezing and melting points (called thermal hysteresis) defines supercooling. Supercooled liquids are unstable—they can turn to ice unpredictably and with little apparent provocation. Mere stirring can be enough. The greater the thermal hysteresis, the greater likelihood that freezing will occur, and the quicker the sample "flashes" into ice, giving off a measurable pulse of heat in the process as the energy of the motion of the liquid molecules is released when they stop their motion after release from the ice crystal lattice.

The absence of antifreeze in the blood of the squirrels, and hence the likelihood of supercooling to as much as 1° to 2°C ought to be risky. A single stray ice crystal in the blood could mean death. Why do squirrels risk it? Why don't they regulate their body temperature 1° to 2°C higher, to avoid supercooling and thus be immune to turning into a block of ice and being killed? Barnes believes the advantage that outweighs the cost is related to energy economy; supercooling to

-2°C would save the squirrels ten times the energy expended by maintaining a body temperature of 0°C (Barnes 1989). The squirrels also have a mechanism that reduces the risks normally associated with supercooling. Generally freezing would start at the coldest point, such as a toe, and Barnes has nucleated (started the freezing process) squirrels' toes and found that the animals were then alerted—they rewarmed quickly before the ice could spread.

Barnes's other remarkable discovery was that the squirrels can and do arouse spontaneously to warm themselves up from a body temperature of less than 0°C, to heat themselves all the way up to their body temperature when active, 37°C. Many other animals can survive cooling to or below 0°C, but none had ever been able to spontaneously arouse unless they were first artificially warmed by being taken to much higher air (and body) temperatures, where the shivering response becomes possible.

Although energy economy helps to explain the squirrel's low body temperature, far from behaving in a manner strictly in accord with just saving energy, the arctic ground squirrels appeared to squander energy by warming up to 37°C from subzero temperatures about a dozen times *throughout* their winter hibernation. Each time they spent about a day being fully warmed and required another to cool back down. These periodic warmings are calculated to cost the animals over half of the fat reserves they had built up during the summer. Why do they bother? It had previously been shown that this behavior is not unique to arctic ground squirrels. Indeed, no mammal hibernator avoids such periodic bouts of normal body temperature during a winter's hibernation. Therefore, given the high energetic costs of warming up and staying warm for a day or so, the behavior seemed most peculiar. It must buy something precious. And this is where Barnes and his colleagues' fourth remarkable discovery comes in. Although still controversial, the hypothesis is that the animals warm up to *sleep!*

Since the early 1950s two different kinds of sleep have been defined. One is "rapid eye movement" (REM) sleep, also called "dreaming sleep" or "deep sleep." The other is defined as "nonrapid eye movement" (NREM) sleep, or "light" or "ordinary sleep." These two types of sleep have been studied in humans by taking voltage measurements from the scalp surface. Brain electrical activity records, called electroencephalograms (EEGS), are plotted against time. In humans, apes, and cats, where the EEG brain-wave patterns have been most studied, there is a progression of very different patterns from awake to four arbitrary stages presumed to represent depth of sleep, with the awake state showing very low voltage amplitude and high frequency (8–13 waves per second) and deep sleep showing a high-voltage amplitude, low-frequency wave pattern. However, when the EEG sleep patterns are followed throughout the night, periods occur in which low amplitude brain-wave patterns reappear that resemble those during wakefulness. It is during these, the REM sleep periods, that the rapid eye movements, and often increased heart rate, changes the breathing pattern, and muscle twitching occurs. The animal is then dreaming.

Animals in hibernation torpor do not show the EEG sleep patterns. Instead, as their body temperature drops and they enter torpor, there is a gradual reduction of voltage until brain electrical activity eventually disappears, as if they were dead. However, although there is no spontaneous brain activity (Lyman and Chatfield 1953), the animals must still *be able* to generate at least some electrical activity in the nervous system, or else they could never arouse.

Teaming up with neurobiologists H. Craig Heller and Serge Daan, Barnes recorded electroencephalograms of squirrels going into and coming out of hibernation. Squirrels entering hibernation showed typical sleep patterns, and then as they cooled down their brain waves disappeared and their EEGs then resembled those in humans who would be considered brain dead. However, once rewarmed by shivering, after having been in hibernation torpor for about a month, the squirrels spent most of the day showing the brain patterns associated with rapid eye movements (REM) during dreaming in humans. Had they heated up to sleep, or to dream? If so, why do they sleep or dream? Why do we? It's one of the large remaining biological mysteries that probably relates in some way to how the brain works to consolidate, edit, delete, and store memory.

It seems ironic that a hibernator has evolved the astounding capacity to arouse from subzero temperatures normally only encountered in winter in order to sleep. If the animals did not need to arouse they could stay torpid until spring and save much energy. Since they don't stay continually torpid despite the obvious energy economy they would experience by doing so, there is apparently some great cost to long torpor or sleep deprivation.

The hibernating arctic ground squirrels may hold keys to the riddle of why we need sleep, and also some medical problems, such as stroke. In hibernating ground squirrels, it is difficult to detect any heartbeat. It is difficult to tell if the animal, with a body temperature less than the freezing point of water, is dead or alive. In a deathlike state of torpor, the animals are cold little balls in which there is only a minute trickle of blood to the brain. In humans when a blood clot or a ruptured blood vessel interrupts blood flow to a part of the brain, there is an almost immediate die-off of brain cells, because our brain cells require a continuous supply of oxygen and glucose that the continuously flowing blood normally supplies. Hypoxia (insufficient oxygen) is the primary deleterious consequence of a stroke, but not in a hibernating squirrel. In the hibernating squirrel's brain there is a metabolic shutdown, so that lack of oxygen and nutrient is less harmful. Do they warm up to oxygenate the brain?

The primary metabolic shutdown in hibernators is due simply to the temperature drop. Human car crash victims who fall unconscious through the ice and whose brain is immediately chilled are also able to survive prolonged hypoxia. But there are also active metabolic processes in the squirrels; brain tissues show suppressed protein synthesis even when warmed to 37°C. A second recent interesting finding is that hibernating squirrels as well as hibernating turtles accumulate five times as much ascorbic acid (vitamin C) in their, as opposed to human, brains. When the squirrels arise from torpor, the vitamin C levels return to normal within hours. It is thought that the vitamin C, a powerful antioxidant, protects the brain from the sudden rush of oxygen that they take in after their long oxygen "fast."

Barnes's interest in hibernation research is motivated by the discovery of basic phenomena, not practical applications. Many others' interest in the hibernating ground squirrel is clinically, rather than intellectually, motivated. They wonder how halting blood supply and hence oxygen and glucose to the brain might be relevant to treating stroke victims, where insufficient oxygen to the brain is the main cause of cell death. They wonder how hibernating animals maintain strong bones despite months of inactivity, why their blood clots so slowly, and why they accumulate huge amounts of vitamin C in their brains and cerebrospinal fluids. I suspect that, with biology becoming ever more applied, research relating to such questions of how to harness hibernation will be increasingly funded in preference to those inquiries motivated purely by intellectual curiosity. That is unfortunate and shortsighted.

The wind whipping through the northern spruce-
fir forest sounds like pounding surf, even as the
thermometer routinely reads -20°C—and sometimes -30°C. I wear wool pants, two sweaters, a
windbreaker, a woolen cap, gloves with liners, wool
stockings, and insulated boots. My fingers become
stiff in minutes when I take off my gloves. Clothing is essential to staying alive, even in the daytime. The cold is no mere abstraction. How do
kinglets, who are out there day and night and who
are no bigger than the end of my thumb, maintain
their body temperature near 43° to 44°C? They
maintain a body temperature some 3°C higher than
that of most birds. For added perspective, that's
6° to 7°C higher than that of a healthy human—
a temperature at which most of us would die of
heat stroke.

Puffed out in its drab-olive garb, a hummingbird-size kinglet looks like a fuzzy little ball. The physics of heating and cooling dictate that small objects cool quickly, since every point within them is close to the surface where heat is lost. The smaller the animal, the proportionally larger is its surface area, which is the drain whereby it loses heat. How can these tiny birds possibly survive even five minutes on a winter day? I know they are there.

After searching in the woods only an hour or so on any morning, I can usually hear the thin *tsees* of a nearby kinglet. I feel wonder, and even after exploring in the woods on hundreds of cold days, I'm still awestruck that anything as small and as dependent on keeping warm can survive. Their calls reassure me that they survived yet another night, and ultimately I must believe my senses more than my rationality.

Golden-crowned kinglets, like humans, are colloquially called "warm-blooded" animals. For us, the problem of surviving winter is similar: how to keep from freezing and have enough energy left after paying heating costs. But for the kinglet, the problem is much more severe, because the greater the difference between body and air temperature, the more energy must be expended to keep warm. Furthermore, the smaller the animal, the higher the energy cost per given body mass.

Good insulation reduces energy costs, but there are limits on the amount of insulation that a small creature can carry. Large mammals, such as musk oxen, wolves, and arctic foxes, put on thick winter coats that insulate them so well that they may not need to shiver on the coldest nights. Snowshoe hares and red squirrels also put on a thicker, though comparably modest, insulating coat in winter (underground hibernators do not have seasonal changes of fur thickness), while still smaller animals generally do not become better insulated in winter.

Birds vary their insulation less by exchanging their garb than by changing how they use it. To conserve heat they fluff out, thereby increasing the depth of the insulating air layer that surrounds them. Foot and leg temperatures stay low, regulated just above the freezing point. (We try to keep our feet warm, and pay a high energy cost for doing so.) When sleeping, kinglets insulate themselves even more by tucking their heads and feet into their inch-thick layer of feathers which, from inside to outside, can maintain the astounding difference between body and air temperature of up to 78°C.

To find out how quickly a fully feathered kinglet loses body heat, I experimentally heated a dead kinglet and then measured its cooling rate. In still air the body temperature of the heated bird dropped 0.037°C per minute for every 1°C difference between body and air temperature. Thus at an air temperature of -34°C a kinglet that maintains a steady 78°C difference between air and body temperature at its normal body temperature of 44°C during activity would have a passive cooling rate of 78 × 0.037°C/min. = 2.89°C/minute. The heat production in a live bird that is required to oppose such cooling can be calculated by multiplying cooling rate by body weight and by the specific heat of flesh (0.8 calories/gram/°C). This calculation showed that a kinglet (with feathers) must expend at least 13 calories per minute to stay warm at -34°C. This is a conservative estimate because a normally active bird would experience moving air, or wind, that would greatly increase the rate of heat loss. Given the above data I could now find out how large a role the kinglet's *feathers* served in insulation—how much energy they save the bird.

I measured the plucked kinglet's cooling rate. My naked kinglet had a 250 percent greater rapid cooling rate than fully feathered ones. That is, a naked bird would experience air temperature at least two to three times colder than a feathered one. Due to its small size, a kinglet would also cool approximately sixty times faster than a naked 150-pound pig. (I was once asked to testify in a murder case— involving a naked human body that was found chilled—on how long the victim could have been dead. Not wishing to experiment on the human body, I used an appropriate animal substitute. I bought a freshly killed, still-warm pig, shaved it, and cooled it to get the appropriate data. Perhaps not coincidentally, I was not called in as a witness when I, an empiricist, told the lawyer my results and conclusions.)

Because of their habit of fluffing out in the winter, kinglets look especially plump. But a plucked one looks like a pink cherry on

Golden-crowned kinglet, close to life-size.

spindly legs. The naked female that I examined weighed 5.43 grams, and she had shed 0.403 gram body feathers and 0.095 gram wing and tail feathers. Thus, she had about 4 times greater feather mass committed to insulation than to flight.

At the time, my naked body weight was 155 pounds. Fully clothed (in a light synthetic L.L. Bean jacket and boots) I came to 166 pounds. Six of these pounds were attributable to the boots alone. The bird had added 7.4 percent to her body weight for insulation, which was about twice the percentage that I had added to insulate myself. However, more than half (55 percent) of the 11 pounds of insulation I wore were designated for my feet. The kinglet, by comparison, used none of its insulation for that purpose.

My little plucked kinglet reminded me of a scaled-down dinosaur. That's not coincidental, as the affinities between some dinosaurs and birds have long been noted. What could not so easily be compared was body temperature, which is, unfortunately, a long-presumed defining difference between birds and reptiles. Since birds are increasingly considered evolutionary descendants of ancient reptiles, it has seemed logical, in evolutionary biology circles, to presume that their ability to regulate body temperature is also a recent, more highly evolved trait. This I dispute. Regulation of a high body temperature is no new thing for us presumably superior warm-blooded vertebrates. It is routine in all large flying insects, ancient animals whose line predates the dinosaurs. Recent discoveries in insects prove that regulation of a high body temperature with the aid of internally generated heat by shivering, which is maintained within the body with the aid of insulation, is not as newly evolved a capacity as was commonly thought. The most parsimonious conclusion, given their size and flight capacity, is that some dragonflies or dragonfly-like insects were endotherms—able to store heat—at least 300 million years ago. And at the present time many other insects of a great taxonomic diversity still regulate body temperature, using internally generated heat, insulation, counter- and alternate-current mechanisms of heat retention and loss, evaporative cooling, basking, and alternating of activity patterns. Others don't regulate their body temperature at all. Whether or not any one does depends almost trivially on body size and lifestyle. What this suggests then is that thermoregulation can be altered through evolution to suit specific circumstances with little respect to ancient ancestors, as such. Insects also tell us that in small animals, endothermy (the process by which internally generated heat is stored to maintain body temperature) is associated with insulation. In many bees and some other insects, the insulation is created from hairlike projections. In moths it is a thick pile created from modified scales instead, and in dragonflies it is a layer of air sacs surrounding the body. In mammals it is fur, and in birds, of course, it is what we call feathers, structures that hold the key to bird body temperature regulation and associated lifestyle. But of course feathers are of *special* importance, because some feathers also serve another, totally different function—they are the birds' passport to flight.

THERE ARE LONG DEBATES in the scientific literature about endothermy in dinosaurs and about the evolution of bird flight. The original rationales were constrained by a multiplicity of simplistic presumptions: Dinosaurs are "reptiles" and present-day reptiles are not insulated and therefore dinosaurs were presumed to be cold-blooded. It was a shock, when in 1861, just two years after Charles Darwin published *On the Origin of Species by Natural Selection*, workmen in a limestone quarry in southern Bavaria discovered a fossil that was clearly a little dinosaur complete with teeth and a long tail, and yet was also a bird, because it showed imprints of feathers exquisitely preserved in the 150-million-year-old Cretaceous limestone. *Archaeopteryx*, or ancient bird, as it was dubbed, was and still is the oldest bird fossil ever found. Having *large* feathers on its front limbs, this ancient bird was likely capable of at least rudimentary flight. But what were the precursors of those feathers? If they were derived from insulation, then even more ancient birds than this one were endothermic. (Unfortunately, subsequently discovered bird fossils have been younger, and nobody has ever found the "missing link" feathers to resolve this question.)

In the 1990s, however, spectacular bird fossils, aged 124 to 147 million years, were uncovered from volcanic ash sediments in China's Liaoning Province. These did not provide direct-line-of-decent type evidence, but they are providing new insights into the evolution of feathers (though not necessarily into the timing and branching points of various possible scenarios of dinosaur-bird relationships). One of the first Liaoning protobirds recovered, named *Confuciornis*, also had the same type of feathers as fully modern birds and it was even more closely related to modern birds than was *Archaeopteryx*. In 1996, however, the same deposits also yielded a small bipedal birdlike theropod dinosaur named *Sinosauropteryx*. Like its relative the Velociraptor dinosaur, *Sinosauropteryx* had sharp teeth, a long tail, and sharp claws. It was a predator with hind limbs built for fast running. Most significantly, though, the body of this *nonbird* had featherlike structures as shown by clear imprint in the fine-grained volcanic ash within which it was exquisitely preserved. Of special note: the feathers on the front limbs of this fossil were totally inadequate to support flight. That is, none of the feathers of this primitive protobird were likely used for flight. *Sinosauropteryx* (and other subsequently discovered dinosaurs with similar feathery structures) strongly suggests, therefore, that feathers for flight originated from insulation. The big question was and is: How can downy insulation evolve into *flight* feathers? I had not given this question much thought but one night in early March 2002, when I was at my cabin in the Maine woods and felt it shake in the wind and heard the pounding of the rain on the roof at night, an idea just popped up out of nowhere, and I present it here.

It had already been a warm winter—by Maine standards, and on the night in question, temperatures were only a degree or two above freezing. So instead of blizzarding, it was pouring rain. After a day of hiking in the woods, my boots had soaked up moisture, and I still had cold feet. My thoughts turned from my feet to Jack London's story of the cheechako who died because of wet feet. The trite and obvious was suddenly steeped with meaning: The kinglet's insulation can do wonders, but if that insulation is wetted then the bird might as well be naked for all of the good it does. That is why a blizzard and subzero temperatures are preferable to being subjected to a cold rain. Almost any rain is a cold, potentially lethal rain. Rain must be a severe test for a small endothermic bird in particular, because the fluffier (and generally the better) the insulation, the more it could act like a sponge to suck out the heat and the life. (Insects simply cool down, and insulation is then irrelevant.) Enduring and surviving wetness must have been a large selective pressure in the evolution of birds when they became continuously or nearly continuously endothermic. How did they evolve to meet this challenge? By the use of *wing* feathers? Is it a coincidence that survival of grouse chicks in Maine depends on rain-free weather in the time before they can fly?

Wing as "raincoat" to protect insulating down.

In a number of the flightless dinosaur fossils we can observe patches of featherlike structures on the arms and shoulders, and these feathers are more tightly organized into flat and regular patterns. They are attached to the posterior surface of the ulna, a bone in the arm, rather than over the rest of the body. However, these already modified feathers could *not* possibly have been used for flight, because they were only 5 to 7 centimeters long. Could they have served instead analogously to how I had seen banana leaves used by other, more latter-day bipedal hikers in the rain in the tropics, who held them over their backs while walking to help shed water? Could such long flat feathers act like a rain guard to retain the insulating capacity of the feathers underneath? Can feathers that one second may be used in flight help channel water off the back in the next? Can the wing feathers act like thatch roofing by helping water slide off along the tight regular patterns of the feather veins?

Down feathers (enlarged).

Intermediate — contour feather.

Water-shedding, shading, and flight feather.

Fundamentally, it's a problem of finding several likely functions of feathers that could have been a bridge between two opposite operations. A prerequisite for any such crossover of function in the bird's evolution of flight is the original

Shielding downy young under mother's wing from rain, cold, and sun.

fluffy downlike insulating feathers of an endothermic dinosaurlike creature must have become even *more* useful as they became *less insulative* (by becoming longer and flattened), long before their utility in supporting flight became possible. Several ideas for the function of stiffened, flattened, and elongated feathers on forelimbs have been proposed. These include using such feathers on the wings as enlarged fingers to help capture prey, to enhance sexual signaling, to boost running speed or maneuverability, and to aid tree-climbing capacity to escape predators.

I'm not claiming that adding another hypothesis—that birds evolved flat feathers on their forelimbs as parasols to reduce wetting of the insulation—adds clarification. It's hard to know now in retrospect, after a hundred million years. On the other hand, it is perhaps almost too obvious that when feather venules are hooked together into a sheetlike structure (as they are in flight feathers and body contour feathers), then they do provide a barrier to both air and water simultaneously. When the forelimbs with such strategically placed feathers are held dorsally over the back, then almost any elaboration of feathers in this direction provides some parasol effect that would help shed water and reduce the wetting of insulative down feathers underneath. If elaborated on further, then eventually such a parasol would need relatively little additional development before becoming

Newly hatched chicken.

Ten days old — now less under hen's wings.

useful in several others of the aforementioned functions, including flight itself. It is analogous to a flying squirrel's fluffy, insulating tail having the hairs arranged into a flat aerofoil. As I thought about the kinglets perched somewhere in the cold with rainwater flowing off their long wing feathers to protect their down underneath, it seemed to me that regardless of whether the parasol theory could resolve a long-standing evolutionary puzzle, it might at least help explain how kinglets survive a stormy night.

I'm searching for kinglets in a stand of red
spruces and balsam firs on the north side of Alder
Stream on a morning in mid-January. It is dark in
these woods even in the summertime because the
dense tree crowns shut out the sunlight. So little
light gets through that only moss grows like a
green velvet carpet on the brown springy humus.
Yellow and purple mushrooms erupt there after
summer rains. In spring the blackburnian, yellow-
rumped, and magnolia warblers sing and build
their nests here.

The warblers are gone now, and it is a different
world. The snow packed onto the dark-needled
branches above me excludes even more light than
it did before, while the snow covering the mossy
ground reveals tracks. Deer have recently been
crossing the brook and their worn path through
this patch of woods continues on to their food in

more open hardwoods beyond. A porcupine has worn a tunnel-like groove through the snow from its shelter under a pile of rocks to a lone hemlock where it feeds at night. Few snowshoe hares cross this area, because there are no twigs within reach for them to eat. The woods seem almost devoid of birds and the gray sky promises even more silence and more snow and the dull crunching of my footsteps muffles the few remaining faint sounds.

Sitting down on the trunk of a fallen balsam fir, I listen for the light but steady tapping of a black-backed woodpecker. Black-backs come here only in the winter, and sometimes I can hear one. They specialize in feeding on bark-beetle larvae found in recently dead spruce and balsam fir trees. Lightly tapping on the bark as if palpating the tree trunk for possible hollow sounds where a larva might reside, these woodpeckers work like a physician tapping a chest for infirmities. No black-backed woodpecker is near today, but I hear a pileated's vigorous hammering. The bird is probably excavating a deep vertical groove into the base of a balsam fir trunk to reach the hibernating carpenter ants deep within the core of the tree. It takes brute force and specialized talents for these insect-eating birds to make a living while remaining in these winter woods. One of the insect-eating birds that seems to be without a visible means of support is the golden-crowned kinglet. This spruce-fir grove is where I go most often to try to find them.

Kinglets have thin voices that are barely audible in the human range of hearing, and they are neither seen nor heard unless you really tune in. Golden-crowned kinglets do not eat the seeds of the various trees that sustain the finches that stay here all winter, nor do they eat the plentiful tree buds, like the squirrels and grouse do. They cannot reach grubs under bark or buried deep in wood. Yet they obviously fuel their raging metabolism to keep warm. They hover at the tips of branches, hop nonstop in dense spruce thickets, and pick at seemingly invisible prey.

The kinglets' tiny bills are suited for gleaning insects from twigs. But what insects could there possibly be about in the winter? How do these gold-crowns manage to find up to three times their own body weight of food each short winter day, as they predictably must to have enough fuel to keep warm? If kinglets are without food for only one or two hours in the daytime, they starve (and freeze) to death. Yet, some of their population clings to the north woods, despite the fifteen- to twenty-hour-long winter nights. Since the birds don't forage at night, and since they have not been observed to cache or store food in the day, what saves them from dying ten times over during the night?

Is that a conversation of barely audible *tsees* in the spruce thicket? No doubt about it. Finally I hear it. And they are coming closer. The sounds are as unobtrusive as a gentle breeze, and they just as easily go unnoticed. There among the thick branches, I finally see one of probably several tiny callers hop and hover hummingbird-like near the end of a branch. A kinglet never holds still for more than a second. Nonstop foraging. Try as I might, I cannot see them actually catching and eating anything, even as I get within three or four feet of them on those rare occasions when they came out of the spruce tops. They feed on prey too tiny for me to see.

What kinglets eat in the winter had long been a mystery, and researchers had relied on speculation to try to decipher what food these insectivorous birds might find in the winter. They reasoned, on the basis of anatomy and behavior, that the birds are springtail (Collembola) specialists. A species of these primitive almost microscopic insects commonly known as "snow fleas" (*Hypogastrura tooliki*, formerly *H. nivicola*) at times pepper the snow in these New England woods. I have seen millions of them gathering in track depressions in the snow, almost blackening the sides and bottom due to the sheer numbers of individuals. They color the snow gray to almost black, depending on their collective numbers. You see the individuals as almost stationary dots that creep about slowly on tiny stubby legs. But when you look closer, you are apt to see some of them hurtle off and disappear in flealike leaps, hence their name of snow fleas. But unlike fleas, they don't jump using their legs. The power for their explosive leaps comes from a muscle-powered spring mechanism attached to a stout tail that is folded under them and then released to strike against the substrate and that tailkick propels them up and away.

Although I have seen almost solid mats of snow fleas on the snow, neither I nor anybody else seems to know where they come from or where they go. Snow fleas appear mysteriously in isolated patches, sometimes within hours after the snow begins to melt. They are on the snow only in the daytime (to absorb sunlight?). Perhaps they burrow directly through the snow, but they also retreat to tree trunks in the evening and radiate out from them in the morning. I have never seen kinglets pay any attention to the snow fleas that are so conspicuous to us on the ground. That is surprising, if kinglets are indeed specifically adapted to forage for springtails, and in Europe Collembola were reported to be their main prey in winter—the manna that provides them the calories and fuels their metabolism to survive the night. Snow fleas appear to be palatable. Bill Barnard, a biologist from Norwich College who studies gray jays in Victory Bog in Vermont, has seen the jays scooping up snow laden with snow fleas like kids eating maple syrup on snow at spring sugaring parties. Do golden-crowns take advantage of this bounty? If so, why had I never seen it happen? Are the springtails up in the trees where the kinglets spend most of their time and where they feed on them unobtrusively? However, as I'll indicate shortly, I doubted that snow fleas spent much time up in the trees.

Kinglets are, as far as I know, the only birds that routinely hover at twig ends to pluck off microscopic mites, aphids, and aphid eggs that probably no other bird could see, nor a human would ever find unless assisted with a hand lens. It would be difficult, but important to know what the birds' lives depend on. Knowing what a bird eats is fundamental if not central to understanding the mysteries of its survival. But knowledge is never without costs to acquire. Sometimes one has to do whatever it takes. To find out what kinglets eat, I had to do the most direct thing. I had to examine kinglets' stomach contents.

I may invite censure for setting a bad example by hunting tiny birds. Nature writer Jim Harrison (1996) warns: "I have noticed lately that hunting, tobacco, and wife-beating are being lumped together in the feel-good quadrant of yuppiedom, that ghastly, fluorescent hell of the professionally sincere that makes one long for the sixties." I'm not insensitive to his observation and need to at least address the hunting rap, given also my upbringing by a father who was a professional bird collector (hunter) for museums. I'll never forget the rhyme he taught me: *"Quale nie ein Tier zum Scherz denn es fühlt wie du den Schmerz"* [Never do something to torture an animal because it feels the pain like you]. He drilled the ethics of hunting into me, and the only time when I ever saw him show anger toward me was when as a preteen in the 1950s I tried to kill a skunk with my slingshot to impress my peers. That was *not* all right. It was disgusting. He felt justified to kill birds for a museum where they would be preserved forever, as some feel justified to eat fish, chicken, or other meat that is digested in hours. Which is more justified? And even if necessary, how do you justify? Those who are familiar with ancient folklore, or are up above the rest of us a moral notch or two, kill "respectfully" by offering prayers or apologies, in the hope that animals will "offer themselves" up to be voluntarily killed. However, it is a sad fact that no animal cares if those who might eat them invent reasons to justify their acts (to make themselves feel good). But if any animals did offer themselves up for the greater good, then none as small as a kinglet would ever consider the value of its meager body to man as sufficient recompense for its own life. So, yes, I killed several kinglets (after getting the appropriate state and federal permits), really only because of curiosity and a hunger for knowledge. And with regrets but no prayers.

I shot the first kinglet at dusk when the bird's stomach would presumably be full. Wanting to maximize the information that it might yield, I took its body temperature as soon as it hit the ground. As mentioned previously, it was an astonishing 44°C (111°F), which is about 2° to 3°C higher than that of most birds. My fingers were rapidly cooling and in danger of numbing, and I then opened the kinglet quickly to see golden yellow fat among the intestines and around its bean-sized gizzard. As I pried open the tiny gizzard to put the contents into a vial of alcohol, I found it filled to capacity. But this gizzard did not contain springtails, as the literature had led me to believe. Not a one. Instead, it contained a total surprise: the partially digested remains (mostly skins) of thirty-nine geometrid ("inchworm") caterpillars. They were a species that neither I nor an entomologist, Ross T. Bell, who made a minute analysis of the stomach contents, could identify. Nobody would ever have predicted caterpillars on trees in the depth of winter, much less in a kinglet's stomach. I would not have been more surprised if I'd have found earthworms.

Geometrid caterpillars of most species are well known to overwinter as pupae, safe from frost in the subnivian zone or deep underground. Nobody had ever reported finding caterpillars on *trees* in the northern winter before. But this bird and subsequent birds, were proof of something new and unexpected.

If birds could find caterpillars, so could we. So, later on in January 1995, after temperatures had again been near -30°F for a few nights, I went out into the woods with four students (Jeremy Cohen, Kristian Omland, Lauri Freedman, and Mike Tatro) on a caterpillar hunt. Having wielded a gun earlier, I now carried a club— a six-foot-long thick trunk from a freshly felled maple tree. The club was heavy, and when I banged trees (up to six inches diameter) as hard as I could, they vibrated from the shock and released a shower of bark and other debris and the tree's crop of overwintering insects onto the snow.

The operation was a big success. I hit fifteen each red spruce, balsam fir, beech, and red and sugar maples and the yield from the seventy-five trees was thirteen tiny geometrid caterpillars (and two small spiders). No springtails. The caterpillars were visible on the snow, but hard to distinguish from needles and debris. They were gray and brown and very small, matching the remains of those I had found in the kinglet's gizzard.

That caterpillars apparently exist on open branches in winter, and that they serve as winter food for kinglets was not known before. But that meant little unless they could be identified. After some checking around I determined that possibly only one man in the world could do it: Douglas Ferguson, at the Systematic Entomology Laboratory of the National Museum of Natural History of the Smithsonian Institution in Washington, D.C. I sent Ferguson some of the larvae I had collected. He examined them and speculated on what they might be, but said that they didn't seem to be a precise match with any he knew. "The only way of identifying such larvae beyond doubt is to rear out adults, which are easy to identify," he said. He had indeed tried to rear the larvae I had sent him but they had all succumbed, probably because they had not fed. There was no way of feeding them until one knew what their food plant is, and there is usually no way of knowing what the food plant is until one knows what the insect's species is!

On another survey—January 1—I banged a total of 224 trees (102 conifers and 122 deciduous trees, 10 species in all). This yielded only eleven caterpillars, all geometrids, and all but four of them were from sugar maples. I decided to try to rear some of these larvae myself. After collecting them at my winter retreat in the Maine woods, I put the frozen caterpillars into glass vials and left them outside the cabin in the snow. Days later when I went back to Vermont, I stuck them into our freezer and, unfortunately, due to other diversions that soon intervened, I forgot about them until July, six months later. When I thawed them, to my great surprise I found five of them still very much alive. Not knowing what to feed them, I put them into a plastic lettuce crisper with sprigs of potential food plants; those found where I had collected them, including spruce, balsam, fir, maple, and beech. On the very next day I was pleasantly surprised when I found five caterpillars, all hungrily feeding. But even though I had collected them on maple twigs, they were feeding on balsam fir needles! So I added more fir twigs, expecting to finally rear moths. No such luck. A few days later all the caterpillars were wrinkled and dead. The cause was a small but by then plump spider that I had inadvertently introduced on a fir twig. Spiders inject digestive juices through their fangs, then suck out the insects' contents, leaving only dry husks. This spider had, apparently, been a very hungry spider.

The kinglet I followed in the maple grove—it was feeding on caterpillars!

The "outbreak" of caterpillars on sugar maples that year was apparently unusual, and it did not go unnoticed by kinglets who otherwise forage exclusively on conifers. On the fourth, fifth, and seventh of January, after cold (-24° to -34°C) and windy nights, I followed two separate pairs of kinglets (for 93 and 75 minutes, respectively), and all four individuals spent the entire time foraging in my young sugar maple grove. I saw them pick off and eat several caterpillars of the same kind that I had also collected. The birds foraged tirelessly, without pause. I timed them at an average of 45 hop-flights per minute, without any apparent change of pace. They bypassed conifers repeatedly (the grove was bordered by balsam, fir, red spruce, and white pines on three sides). Given the windy and cold nights that must have produced a chill factor of near -50°C, I marveled not only that they were alive, but that they switched from spruce/fir to maple, apparently having learned to associate food with specific trees.

Each bird species, like every organism on earth, feels most at home in the specific environment to which it has been tailored by natural selection and instinctively seeks that environment and avoids others. For kinglets, that environment is dense coniferous forest, as for us it is probably the open savannah where we can have long open vistas, scattered trees, and water. However, we are much more flexible than kinglets and can readjust; although I find the clearing around the cabin with the view toward the mountains and the lake aesthetically appealing, I also love the forest. The wonder is that although most birds are strictly programmed to remain in specific environments, the kinglet can escape that programming when it has to. It pursues caterpillars even when they occur outside its normal bounds.

In still a third winter, 1999–2000, I joined a third pair of students, Joshua Rosenberg and Jonathon Taylor, and took up tree banging to again sample winter insects on trees to find out where most of them are, and what they might be. We sampled six tree species (thirty trees of each). As before, we clubbed trees with some force to see what would fall off onto the snow. (Not all caterpillars are knocked off by this method. Some caterpillars are attached by a thin silk anchor to the twigs. Sometimes they were left hanging from this safety line, and when they thawed out they used that lifeline to crawl back onto the branch.) Most of the sample (87 percent) was geometrid caterpillars, for a total of 80 larvae from the total of 180 trees. This year the number of larvae from sugar maples (13), was not significantly different from red spruce (11), beech (19), although the number was significantly higher on pine (30) and lower on balsam fir (2) and red maple (5). In this winter and all subsequent winters, we did not again find the kinglets foraging in the sugar maples, making their behavior of the previous year all the more remarkable.

Caterpillar on beech twig.

As before, I again brought some of the caterpillars home. All had survived being frozen, and all looked healthy when I took them out of our freezer in May. As before, I offered the larvae maple and fir foliage (carefully checking for any stray spiders this time). But the caterpillars refused to eat, except one lone larva (a very thin dark animal) that fed on the fir. It pupated on June 22. Unfortunately, I made another mistake. I had left it on the windowsill in a small jar, and moisture that had condensed on the glass wall drowned the pupa. I was, of course, very disappointed to finally almost have a moth, and then lose it.

Getting close to an identification (I hoped), I tried again the next year. I reared a number of the larvae that I again collected on sugar maple but fed on balsam fir, to adulthood. Beautifully gray-mottled moths variously marked with subtle browns and cream emerged, and Ferguson identified them as the well-known, one-spotted variant, *Hypagyrtis unipunctata*. The food plant of this species is reported to be extremely variable, including alders, willows, birches, oaks, and balsam firs. Charles V. Covell's book *Eastern Moths* describes the variant as "extremely variable sexually, geographically and seasonally." It had not previously been known where the larvae overwinter.

The "variant" moth reared from winter caterpillars.

It had been difficult to identify the caterpillars. We were poor taxonomists, even of trees, as my exam of the students proved. But the kinglets had apparently learned what I'd been trying to teach my students, generally less successfully. Are they intelligent? Kinglets, being small, cannot have large absolute brain size. Nevertheless, on a per-body-weight basis, their brain mass is massive. It accounts for an incredible 6.8 percent of their whole body weight (as opposed to ours of about 1.9 percent). Thus although a kinglet's total brain mass does not amount to much in absolute terms, it does represent an enormous commitment to neurons given the size of the bird.

A brain is metabolically expensive, and kinglets live on an energy edge in winter and have no energy to spare. In humans the 1.9 percent of body tissue that is devoted to brain mass reputedly accounts for 20 percent of our energy drain. There is much debate about what *our* energy base could have been in order for us to evolve to support such a large brain and why we should have it.

In kinglets, the energy drain of the brain could be triple ours, and we now have the answer to the energy source question: caterpillars. Turtles are successful because the brain drain has been reduced to a minimum—a barely enlarged bump on the nerve cord—helping them to survive up to a year without food. When all is said and done, chances are we'll never know why the kinglets' brain size scales large. However, we can be reasonably sure that if their brain is not part of the problem of energy balance, then it is likely to be part of the solution. We don't know how kinglets decide what to eat in the winter, what else they eat, and hence what their flexibility is. But finding out they eat moth caterpillars in the winter is not only a satisfying accomplishment for all of us who took part, but it is also the discovery of a link in their survival. To care for the welfare of kinglets, it is necessary to care for moths.

The idea that birds hibernate started with a belief that the swallows that skim over the ponds in the fall spend the winter in the mud under the ice. After we learned about intercontinental migration, a more remarkable phenomenon, the first idea seemed so ridiculous that any mention even of hibernation in birds was automatically deemed nutty. Nevertheless, torpid birds were eventually found, and by highly reputable observers, including Nobel laureate Konrad Lorenz. When W. L. McAtee found a torpid, seemingly almost dead chimney swift, a normally migratory species, in mid-October 1902 (in Indiana), which quickly recovered in a warm room, he said: "That experience gave me an interest in avian torpidity and stimulated the collection of references on the subject" that he summarized. Among these references are accounts of torpid swifts and other "swallows" that were found in hollow trees, stovepipes, and rock crevices and clefts during the winter, after most of the population had disappeared. Most of these early accounts seem credible, aside from the likelihood that some of the "swallows" may have been swifts (since the terms were then often used interchangeably). However, in no case were the durations of torpor ascertained to see if the term "hibernation" was justified. All that is, except one set of observations made in North America fifty-five years ago on the poorwill, a bird I've met only in the literature.

The Birds of America (1917) is a large multiedited and handsome book illustrated with sketches, photographs, and 106 full-page paintings by the bird artist Louis Agassiz Fuertes, who went, I think, far beyond his famous predecessor John James Audubon in his handsome and lifelike bird paintings. I got it for a Christmas present when I was eleven years old from a neighbor, Mary Gilmore, who never knew how much joy she gave. It is, in my opinion, the best book of American birds ever produced, and its editor in chief, T. Gilbert Pearson, had this to say about the poorwill (*Phalaenoptilus nuttallii*, the western relative of the whippoorwill that I used to hear in Maine then):

Torpid poorwill (drawn from a photo in Jaeger).

"I first heard the song of the poor-will in a wild canyon in the mountains of New Mexico. In company with Charles F. Lumis, the archeologist, I was camping in the long-silent homes of the cliff dwellers, high up on the white tufa walls of the haunted cliffs of Tyu-on-yi. It was a quiet summer night with the moon shining in great brilliancy. The surroundings were most impressive, and when the sudden cry of *poor-will, poor-will,* was borne on the air from across the canyon, it was as if a voice from the spirit-land had spoken." He went on to say that this approximately seven-inch bird is strictly nocturnal, and like others of the Caprimulgidae, or "goat suckers," it catches only flying insects. He presumed therefore, reasonably enough, that "these birds retire southward when winter appears."

We gauge what we think is possible by what we know from experience, and our acceptance of scientific insights, in particular, is incremental, gained one experience at a time. Just as there was still much to experience about poorwills at the turn of the century, there is still much unknown about kinglets even now. One of those experiences that considerably stretched the physiological limits of what was thought possible for any bird after Pearson produced his book was an amazing revelation by Edmund C. Jaeger made in the winter of 1946–1947 in the Chuckwalla Mountains of California's Colorado Desert (Jaeger 1948). The poorwill, which the Hopi Indians called the "sleeping one," apparently hibernates rather than migrating. The Navajo are also familiar with the birds, and when Jaeger asked a Navajo boy whom he knew, "Where do they stay in the winter?" the boy answered, "Up in the rocks."

Sure enough, on a granite ledge in a secluded crypt lay what appeared to be a dead poorwill. But appearances are often deceiving. After Jaeger picked up the bird, it stirred in his hand, came fully back to life, and flew away. The next year in late November 1947, Jaeger again found a comatose poorwill—maybe the same poorwill—back at the same spot. He checked on it at weekly intervals and it always appeared to be dead. But when he last visited it that winter, on February 22, 1948, the poorwill immediately flew out of his hand when it was removed from its place of hiding. According to Jaeger's calculations, the bird was presumed to be in a comalike torpor for about eighty-five days, the time when there were no or very few flying insects in the Colorado Desert. On the evening of February 22, Jaeger first noticed many insects flying into his campfire and into the beams of his car headlights.

On the five mornings during the preceding eighty-five days, Jaeger had measured the bird's internal temperature by inserting a thermo-probe into its cloaca. All his readings showed the bird's internal temperature hovered near that of the air temperature, as is usually the case when an animal is dead. He could detect no heartbeat with an aid of a medical stethoscope and saw no breathing movements of the chest. No moisture collected on a cold mirror held in front of the bird's nostrils. A flashlight shined a full minute into the bird's right eye (which was almost completely open) failed to elicit any response whatsoever. Jaeger concluded: "I take it as evidence that the bird was in an exceedingly low state of metabolism, akin, if not actually identical with hibernation, as seen in mammals" (Jaeger 1949, p. 106).

To determine whether the bird returned every winter to the same spot, Jaeger banded it with a U.S. Fish and Wildlife Service aluminum cuff. To his great satisfaction, the bird returned to its hibernaculum on the granite cliff in the winter of 1948–1949. It survived a severe November storm of sleet and hail that left a layer of ice on the ground for a day afterward and air temperatures that remained near 0°C.

Given what the Hopi and Navajo knew already, we can't say without qualification that Jaeger was making a discovery. But his reports surprised physiological ecologists perhaps as much as if he had verified the old fable that swallows hibernate in the mud. Jaeger's two papers begat a flurry of lab studies: fifteen laboratory studies of the poorwill and related species have since appeared in the scientific literature. These reports extend, and perhaps require some reinterpretation (but not much) of, Jaeger's original paper. They confirmed that the poorwill's body temperature in torpor indeed becomes essentially identical with air temperature (Howell and Bartholomew 1959). The torpid birds can spontaneously arouse from body and air temperatures as low as 6°C, although when doing so the process can take several hours (Ligon 1970). At such low air temperatures, however, the physiological capability of arousing is seldom used (Howell and Bartholomew 1959), presumably because it costs the bird too much time and energy (and probably also because it brings them little since there are then no flying insects to catch).

The poorwill, although unable to arouse from temperatures near 0°C, can nevertheless *survive* such low body or air temperatures (Ligon 1970). Thus, there is no problem explaining the survival of the particular bird Jaeger observed in an ice storm at temperatures at or below 0°C. But I doubt that his birds spent eighty-five days in continuous torpor. Captive poorwills regularly enter torpor at night, but don't remain in that state for more than four days at a time (Marshall 1955). Since Jaeger always measured body temperature of his bird between 10:20 A.M. and 11:30 A.M., it seems not unlikely to me that his poorwill could have warmed up to forage on some warm nights, to then return to its same perch and resume torpor by the next morning. Other caprimulgids, like the nighthawk, which is migratory, are much more reluctant to become torpid at night (Lasiewski and Dawson 1964), probably because they escape cold weather instead.

As would be revealed in the avalanche of studies following Jaeger, such patterns as those in the poorwill are only a slight modification on those later documented in innumerable other birds. The main feature that makes the poorwill remarkable is that it sometimes stays torpid for days at a time without rewarming, thereby showing that physiologically there is no distinction between hibernation and daily torpor.

THE SMALLER THE SIZE, the greater the physical tendency for rapid cooling and the greater the energetic advantage to staying cool. A turkey-sized animal is not likely to let itself become cold enough for torpor at night, but the smallest birds and mammals, weighing as little as about 3 grams, including hummingbirds, some bats, and some mice and endothermic insects, do so routinely, actively, and as an adaptive response. They are the daily hibernators. By stopping to shiver at the cessation of daily (or nightly) activity bouts, some may cool down to ambient temperature in minutes. They may then be picked up and seem dead or dying, and one often assumes such. However, when ready to resume activity a half day (or night) later, they shiver and heat up explosively. In minutes they are as active as before. Small perching birds weighing more than 50 grams do not generally go fully torpid at night, but they save at least some energy by reducing body temperature a few degrees Celsius.

The torpor option for conserving energy was little known until the 1950s, when Jon Steen (1958) in Norway examined the metabolism of titmice and five species of common finches captured one winter just outside his lab near Oslo. At night the birds reduced their heat loss by balling up—tucking their heads into their back feathers—but they also reduced heat production by lowered body temperature at night. However, when plenty of food was available, the birds maintained normal body temperature at night, shivering often the whole time, while sound asleep.

Torpor of one degree or another is a well-known strategy for conserving energy of many of our winter birds. No bird in North America has been as well studied in that regard, and is as familiar to many people, as is the black-capped chickadee (*Parus atricapillus*). This 10- to 12-gram bird takes sunflower seeds from bird feeders all winter, and no walk in the snow-filled winter woods is complete without at least one run-in with a family flock of these small, tame, and inquisitive birds as they search for food. Day in, day out they are active, no matter how cold the weather. In Alaska the birds show a peak of food-hoarding activity in November (Kessel 1976), similar to other tits (Nakamura and Wako 1988). Nevertheless, how they maintain an energy balance during long cold night fasts was a mystery until they were investigated in the early 1970s by Susan Budd Chaplin at Cornell University.

Chaplin's study of chickadees in the Cayuga Lake Basin in New York focused on the most critical time for cold survival: winter nights. She began by determining energy expenditure of the birds as they regulated their normal day-active body temperature of 42°C at air temperatures as low as 0°C and as high as 30°C. She recorded the chickadees' rates of energy expenditure (oxygen consumption) when they were placed into constant temperature in a sealed chamber in the lab. As in all other homeotherms (animals that regulate a high and constant body temperature), metabolic rates were predictably high at low air temperatures, to make up for increased rates of heat loss. Given the hours of darkness per night, Chaplin could calculate how many calories of energy reserves a bird might require to enable it to remain fully heated over the duration of a night, and then she compared to see how that number correlated with the calories stored in the body's fat reserves.

Fat has more than twice the caloric content per unit weight as carbohydrates (such as sugar and glycogen). Fat is therefore the fuel of choice for long-distance travel and for other long durations of exercise, such as all-night nocturnal shivering of sleeping birds (Marsh and Dawson 1982). The chickadees' fat reserves were determined by collecting birds in the woods (by shotgun) in the evening and again in the early morning, then chemically extracting their fat contents to see how much energy they used during the night. Body fat in the evening measured 7 percent and only 3 percent in the early morning. That is, the birds fattened up throughout the day and then burned off their fat to produce heat to keep warm during the night. Winter birds had higher metabolism (Rising and Hudson 1974) and generally maintained twice the fat content of fall and spring birds. Indeed, the ability in any birds to put on body fat in significant amounts is generally restricted to only those that need it for migration or surviving cold while experiencing food shortages. Birds with a dependable food supply and resident in a moderate climate show little or no fattening (Blem 1975), presumably because there are costs of being fat; it is better for them to be lean unless there are compelling reasons to the contrary.

Given Susan Chaplin's figures, chickadees are already close to an energy edge at 0°C, far from the lowest temperatures they might encounter during any winter night. Unlike some seed-eating birds, who usually have more food calories available and who put on a lot more fat overnight (Reinertsen and Haftorn 1986), Chaplin's chickadees did not have sufficient caloric reserves in fat to make it through a night at 0°C, if they continued to regulate the same body temperature at night as during the day. However, she discovered that, unlike the seed-eaters, they stretched their fat reserves by lowering their body temperature to 30° from 32°C, that is, to 10° to 12°C below the 42°C of normally regulated daytime body temperature. This readjustment of their body temperature set-point was sufficient to make their fat reserves last the night, despite vigorous shivering during most of the time they were sound asleep. Nevertheless, even with the caloric savings derived from hypothermia, the chickadees' fat reserves in the morning were insufficient to last them through another day and night, such as could occur during a severe blizzard. To survive such commonly occurring emergencies or temperatures much lower than 0°C would require them to have special shelter at night where air temperatures are higher and convective cooling minimized and considerable energy would be saved (Buttemer et al. 1987).

Chickadees do not build winter sleeping shelters, but like other Paridae they show great flexibility in choice of roost sites for overnighting (Perrins 1976; Pitts 1976). Black-capped chickadees may sleep in almost any tight cranny or cavity (as can sometimes be deduced from their bent tail feathers in the morning); in dense vegetation such as vines; in conifers; and possibly in snow. Siberian tits have even been reported to dig 8-inch-long tunnels into the snow for overnighting (Zonov 1967). Although black-capped chickadees have not been reported to huddle at night or dig into snow, many of their relatives do (Smith 1991, p. 246).

Chickadee fluffing its feathers on a cold winter day. Tail feathers are bent from overnighting in cramped quarters to escape cold. (Drawn from photograph by David G. Aden.)

One of the chickadees' remarkable winter adaptations is their plumage, which is denser than that of other birds their size (Chaplin 1982; Hill, Beaver, and Veghte 1980). Heat loss is mainly from the area around the eye and bill, and when the birds fluff out and then ball up to sleep, they are reducing specifically that area of heat loss by tucking their heads under their scapular (shoulder) feathers of the wing.

The fact that chickadees were cutting it close, though, even at modest winter air temperatures, only deepens the mystery of how golden-crowned kinglets survive. They are half the body size of chickadees and experience at some times double the temperature extremes of Chapin's study birds. Do kinglets overnight without much larger fat reserves?

Charles R. Blem and John F. Pagels from Virginia Commonwealth University have provided the only data to help answer that question. In January–February 1983 they collected (shot) kinglets in Virginia throughout the day. Like the chickadees, the diminutive golden-crowned kinglets increased their fat stores during the day, from about 0.25 gram at 8 A.M. to about 0.60 gram at 5 P.M. Despite the kinglets weighing only half as much as the chickadees, these amounts of fat are nearly the same in absolute terms as the chickadees'. Thus, relative to body size, the kinglets put on twice as much fat per day as the chickadees. Nevertheless, even these fat reserves already seem low at modest air temperatures of 0°C for the northern winter nights of fifteen hours; Blem and Pagels calculated that a kinglet in such conditions would require approximately twice the calories contained in their maximum fat reserves to last the night, if they regulated their day-active body temperatures. The mystery of how they manage was not, and still is not, answered. Reductions in body temperature are likely although the one study that examined this possibility (in captive birds) found *no* hypothermia. I suspect that in the wild, at -30°C, and in fifteen-hour nights, they *must* become hypothermic.

The relevant question is *how* hyperthermic? No body temperature measurements of kinglets in the wild under severe (or any) winter conditions (such as -30°C and wind) are available, so we have no definite answer. The birds would predictably have a great and most urgent need of hypothermia for *energetic* economy at -30°C, but such regularly encountered temperatures would pose a great risk of freezing to death in birds that become too hypothermic. Birds that get too cold could become unable to respond. Not being able to shiver they might then quickly turn to ice; cooling down risks losing physiological control for being able to generate heat. The trick is to be able, like the arctic ground squirrels, to achieve a physiological state that is technically close to death, while retaining the ability to respond and come back to life on demand. Minihibernation overnight is a good strategy, but only if temperatures in the morning are high enough to allow the animal to passively heat up to the point where shivering is again possible and the bird is able to warm up quickly. Endothermic insects face the same problems as small endothermic vertebrates, but more acutely.

Consider, for example, the tomato sphinx moth (*Manduca sexta*), whose large familiar green larvae feed on tomato and other solanaceous plants. The nocturnal adult regulates nearly the same body temperature as does a hummingbird while in flight. After flight at, say 15°C, the moth immediately cools and within a minute or two it is torpid. In the evening, if air temperature is 30°C, it needs to shiver for less than a minute to be flight-ready again. But if air temperature is 15°C it must shiver for several minutes. If air temperature is only 5°C lower, however, then the animal is incapable of warming up at all. It would remain in torpor and assuredly freeze to death if temperatures were then lowered to below 0°C. Normally, however, summer-active moths are never subjected to such low temperatures, so they need no defense mechanism to escape death by freezing. Similarly, bats can afford to enter hypothermia to save energy, when they are within the safety net of a cool but not too cold cave. They can slip into torpor, secure that they won't turn into a block of ice, as long as they overwinter in deep caves where temperatures don't go below the freezing point of their tissues.

The owlet moths of the subfamily Cuculiinae that are common in New England, face the brunt of the problem of potential lethal freezing. To escape predators (bats) they are active in the winter. Their flight muscles are amazingly cold-tolerant, and they can shiver and warm up even from temperatures as low as 0°C, but they will freeze solid at near -10°C. Nevertheless, they don't shiver to prevent themselves from cooling to lethal temperatures when subjected to temperatures approaching 0°C. Instead they seek refuge under insulating leaves and snow to avoid getting that cold. Neither the moths nor most ground squirrels, except arctic ground squirrels, are in any real danger because in the microhabitat where they hibernate they are sequestered from very low temperatures and so no alarm response to dangerously low body temperature has evolved in them. In general, few birds find secure shelter from the cold such as that afforded by

One of the noctuid winter moths at rest on a beech twig.

deep caves or underground burrows. Winter birds may face temperature drops of from well above freezing to -30°C or colder over the course of a single night, and they simply can't afford to relinquish body temperature control. Ironically, hummingbirds provide a conspicuous model of an adaptive response that applies to many birds in winter.

Nocturnal hypothermia is common in hummingbirds because of their small size, although if energy supplies are available and temperatures are not too low, the birds don't have to resort to this option. In the black-chinned *Archilochus alexandri* and the Rivoli's *Eugenes fulgens,* torpor is used only in an energy emergency (Hainsworth, Collins, and Wolf 1977). Similarly, in the broad-tailed hummingbird (*Selasphorus platycercus*), which successfully rears its young in the energetically near-marginal conditions of the Rocky Mountains, can go torpid even when incubating on the nest if energy crises result from rainstorms and low nighttime temperatures. In other hummingbirds, torpor occurs even in very fat birds. For them it serves as a mechanism to conserve their energy resources needed for migration (Carpenter and Hixon 1988). The Anna hummingbird (*Calypte anna*) found from northern California to Baja California, regulates its daily energy budget less by nocturnal torpor than by daily gain of energy stores, increasing its body mass by over 16 percent during the course of the day.

There is obvious benefit of torpor, provided the risk of losing physiological control in an environment where temperatures dip too low, is not too great. Some hummingbirds are unable to respond to lowering temperature by shivering if they cool down to 20°C (Withers 1977). These are species (*Calypte anna* and *Selasphorus sasin*) living where they don't encounter temperatures lower than 20°C (southern California). Others, from colder mountain environments, regulate not only a high body temperature when active but also a low one when in torpor (Wolf and Hainsworth 1972). Two other hummingbird species (*Panterpe insignis* and *Eugenes fulgens*), from the high cool mountains of Costa Rica and western Panama, not only regulate but are capable of spontaneous arousal from body temperatures of as low as 10° to 12°C. As already mentioned, the arctic ground squirrel, a hibernator, was later shown to do the same at even much more impressively low temperatures. Some hamsters (Lyman 1948) and pocket mice (Tucker 1965) have also been observed to first allow themselves to become torpid but then retain the ability to resist cooling below a specific, much lower body temperature threshold.

One cannot predict what golden-crowned kinglets do in any specific area and under specific conditions. All we can be reasonably sure of is that they likely engage in *some* torpor, but very deep torpor is probably not an option. A kinglet in a windy winter night at -30°C would have to remain ever-alert. If it should stop shivering for several minutes, it would quickly freeze as solid as a teaspoon full of water.

There is, of course, a way out of the kinglet's quandary, and that is to find a microenvironment, such as a verdin's nest stuffed full of body warmers. However, such options are limited in the north woods. This habitat contains neither verdins nor birds who build nests like them. Kinglets are too large to burrow unnoticed into the feathers of an owl, as do hippoboscine flies in winter (and summer). And they are too small and frail to burrow into the subnivian zone under the snow like grouse do. They obviously can't avoid freezing to death by diving under ice-crusted waters. And yet I've seen dippers (relatives of wrens) in midwinter in Yellowstone Park jump into the icy swift water of the Lamar River, disappear from view, and later pop up along the edge of the ice.

I'm not implying that I think even for one second that dippers, or kinglets, can stay down and hide in some crevice under the bank like a frog or a trout. There are lots of obvious reasons why they don't, but why couldn't they have evolved that ability? Much that animals have evolved to do would have seemed impossible to us, if experience had not taught us otherwise. And few inhabitants of the winter world have evolved more ingeniously, more bizarrely even, than turtles, frogs, and insects.

Snapping turtles *are air-breathing reptiles*

that lay buried in the mud at the bottom of frozen-
over ponds for six months of the year, without ever
once coming up for air. Except to lay their eggs,
these turtles rarely emerge from their watery world.
I recall one I saw in Pease Pond near Dryden,
Maine, when I was a kid fishing for perch. The turtle was as big as a washtub, and it was swimming
slowly under our rowboat like some prehistoric
monster. In my imagination then and memories
now, it could well have been a plesiosaur.

There are snapping turtles in the beaver pond
near my house in Vermont now. In early June the
egg-laden females make their short migration up
and out from the beaver bogs to their traditional
nest sites. A by-now-familiar foot-long snapping
turtle has chosen a sunlit patch of gravel alongside
our neighbor's driveway. In June she scoops out a

cavity at that spot with her hind feet and deposits about a dozen white leathery eggs. Then, after covering them, she lumbers on back down to the bog. In early September the hatchlings dig to the surface, cross the road through the woods, and they too slip into the bog. There they bury themselves in the mud and remain until spring.

Turtles are measured in superlatives, from the Galapagos turtle that lives 150 years to the oceanic leatherback that weighs up to fifteen hundred pounds. Turtles have been around and have changed little since the Triassic ages of about 200 million years ago. They shared the earth with the dinosaurs for well over a hundred million years. They have adapted to deserts, oceans, and cold climates. To me they are the most interesting and attractive of reptiles, and I find baby turtles especially appealing. That even includes the baby snapping turtles, whose tails are longer than their bodies and who look like miniature alligators. Unlike the young of birds and mammals, turtles seem complete and self-sufficient replicas of the adults.

The common snapping turtle (*Chelydra serpentina*) can grow to be three feet long, snout to end of tail, and weigh up to fifty pounds. When they are out of the water, snappers live up to their name. They will lunge at you, and can reputedly snap a broomstick (probably an exaggeration). Still, you don't mess with them. One early morning in September 2000, as I saw the first white frost on some of the grass and the purple New England asters were just starting to flower, feeding the migrating monarch butterflies coming by daily in droves, I heard a splashing down on the beaver pond. Geese? The splashing continued. A wading moose? I rushed down and peeked through the thick foliage. Mallards—about twenty of them—were flying around the pond in alarm, landing in scattered groups. No geese and no moose were in sight, but on the other side of the pond, close to shore, I saw a steady splashing-churning of the water. I studied this strange phenomenon through my binoculars, without getting any clues. I would have to get closer, so off I ran through the woods and through ankle-deep sedgy shallows, then onto the beaver dam where numerous deer had also crossed recently, judging by the fresh tracks in the new mud laid down by the beavers.

Once I got near the commotion, I saw that it was a duck helplessly flapping in place. I waded out through the muddy water beyond sedge hummocks, till I neared the duck. It flailed harder, dove and disappeared totally from my sight in the now quite muddy water, but reappeared in seconds. Finally I grabbed it, and it then ceased all movement. As I suspected, the duck was attached to something solid. A log? I lifted the duck a bit higher, exposing yellow-pinkish legs and feet. There, attached to one foot, was an object the size of a quart jar, coming to a triangle near the front. It was algae-covered—except for the eyes. The snapping turtle had a solid grip on the duck's right foot.

I presume the duck was lured within striking range because the turtle's back looked like a moss-covered rock for a convenient perch. Although this may not be the designated hunting strategy of a couch potato, it may approximate it. This turtle's close relative, the alligator snapper (*Macroclemys temmincki*), clocking in at a record weight of two

Common snapping turtle.

hundred pounds, is probably the ultimate low-energy investment hunting specialist. Lying on the bottom with its mouth open, this turtle just wiggles its pink, wormlike tongue and lures its unsuspecting prey directly into its mouth. Talk about efficiency.

I started walking backward toward the shore, pulling the duck with the turtle attached to its foot. The turtle would not let go. I looked into its hard yellow eye with its starred little black iris—the eye that had, scarcely unchanged, seen the dinosaurs come and go.

How could I get the turtle off? I wondered—because I was not going to let this creature finish drowning the duck. That might take hours in this shallow water. I had no weapon at hand, and contemplated slowly dragging the monster to shore and finding a stick to beat it over the head. Maybe it would then let go.

As I was thus slowly maneuvering the pair toward shore, the turtle finally obliged and let go. Now I held a duck, which still had not moved and had remained totally silent. The web between its toes was badly torn. But this injury would heal. I threw the duck into the air. It quacked a few times, and flew off.

Meanwhile the turtle slowly, ponderously, moved on the bottom out of the stirred mud into less murky water. I reached down and grabbed its long tail. I don't know why. The engagement it had had with the duck was long enough. Ours somehow, wasn't yet. But what to do? I lifted it—hefted it—but perhaps wisely not all the way out of the water, even if I could have, which wasn't a sure thing. But I lifted it high enough to see its pale yellow underside—of thick neck and belly. What could I do? Nothing. I let it go—it resumed its slow lumbering journey away from the shore, toward deeper water. Perhaps I should have felt guilty—I may have deprived it of its last meal before it would fast for six months while stuck in the mud.

A year later, as the latest crop of baby common snapping turtles were digging themselves out of their nest in the gravel of my neighbor's yard also to enter hibernation quarters in the pond, I scooped three of them up and put them into an aquarium along with min-nows, whirligig (Gyrinid) water beetles, tadpoles, and one giant predaceous (Dytiscid) water beetle larva. The hatchlings seemed lively enough, but they refused all food. Instead, one *became* food: within a day the water beetle larva killed it by clamping its hollow pincers into the turtle's neck and injecting digestive juices. I removed the predaceous beetle larva while it was sucking up its meal, and then left the aquarium outside.

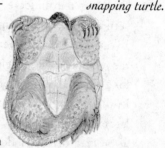

Hibernating baby snapping turtle.

By December the aquarium had ac-quired a thick layer of ice. The ice did not slow down the minnows or the beetles, which both remained as quick as before, but the turtles settled to the bottom and looked dead. I presumed that was normal, for a turtle. Finally in late December, I brought the aquarium inside and when I removed the remaining ice and pulled up the turtles, they still seemed stone-dead. However, as soon as they warmed up they became as lively as they'd been before. They were now likely smaller (lighter) since they had not yet eaten anything in the three months since hatching out. What had changed was that they were now voracious feeders, and they could only now start to grow.

Perhaps hatchling snapping turtles don't feed until after they have hibernated. If so, they are among the few creatures that normally start their life with an eight-month fast. At near 0°C, their metabo-lism shuts down and helps them conserve energy and/or reduce their need for oxygen. In contrast, northern fish compensate and adapt their metabolic machinery to be *active* despite the otherwise normally depressed metabolism due to low temperature. (Summer-active minnows that I put into ice water went belly up in seconds.) In fish, that temperature compensation involves activating new enzymes (isoenzymes) that perform the same function but that operate at lower temperatures where the previous enzymes would normally shut down.

A PROBLEM THE TURTLES and other winter water dwellers face that is closely related to food or energy supply, is oxygen supply. All turtles breathe air with lungs, but many species spend the entire winter without the opportunity to take a single breath of air, since they remain locked in under the ice. I have on rare occasions seen a turtle rowing along in slow motion just under the clear ice of an early October or November freeze-up. The turtles' longest dives of the year are then just beginning, or may have already been in progress for a month. The duration of a turtle's dive is determined by its ability to get oxygen from the environment and by its ability to accumulate an oxygen debt in its tissues. For dives of a winter duration in an *air breather,* that precludes engaging in vigorous, or almost any exercise, which snappers manage nicely.

The first sheet of clear transparent ice that covers the pond still allows sunlight to penetrate to the pond plants, and their photosynthesis still produces oxygen at low rates. The oxygen they give off dissolves in the water and is sealed in. As snow later covers the ice, less light penetrates to power the plants and their vegetative portions die. Now they start to rot, and to remove oxygen from the water.

Low water-temperature is an advantage to many organisms, because cold water absorbs and retains more oxygen than warm water. Most of the active water animals have gills to take in that oxygen. The aquatic insect larvae have them—dragonflies, damselflies, caddisflies, stoneflies, mayflies—as do the overwintering tadpoles of green frogs, bullfrogs, and some others. However, none of the few *adult* insects living underwater have gills, even those which are fully aquatic. That's probably because the adults have only secondarily invaded the water, and their air-breathing mechanisms have continued with special adjustments added for life in the water. As adults, they were evolutionarily locked in to air-breathing.

Diving beetle adults and aquatic bugs carry air down with them. The predacious diving beetle *Dytiscus*, which captures tadpoles and small fish (and one of whose larvae killed my snapping turtle), carries a bubble of air hidden under its wing covers that it may expose to the water so that oxygen can diffuse in. Some other beetles, *Hydrophilidae*, and the back-swimming bug *Notonecta*, have their ventral body surface covered with a thin film of air (called a plastron) that shines silvery in the water. Like the Dytiscid's air bubble, this air layer is connected to their air-using tracheal system, and oxygen used up from the air film attached to their bodies is replaced by oxygen from the water passively following the concentration gradient. As a result they can stay active even as ice covers the water.

The water under the ice provides an ideal environment for the animals that can breathe there. It is the one assured refuge from freezing, and many predators are excluded. For centuries it was presumed that birds spent the winter there. Not knowing much about bird biology and evolutionary constraints, it could long ago have seemed logical to observers that the swallows in the fall that skimmed closely over the water surface would spend the winter in the mud under the ice, since frogs, salamanders, and the myriad insects that emerge as adults from the water in the spring fly off and often live far from water. Of course, birds don't hibernate in the mud, and it is not because an impossible physical hurdle stands in the way for evolution of such capacity. The major problem is probably evolutionary inertia. You can't convert a jet plane to a prop plane, or vice versa. But what you can do is improve each, up to a point. As with the adult air-breathing insects, birds are historically locked into air-breathing. You can't just make them water-breathing. And turtles?

Present-day turtles include the already-aquatic species that are evolutionarily predisposed or preadapted for life in the water. Even now, turtles are arguably the world's best divers, and a winter's hibernation under the ice is a prolonged dive—one that may extend to over six months of the year. Durations depend on the species and the physical characteristics of the specific hibernation site chosen.

In one study by Gordon R. Ultsch and colleagues (2000), map turtles (*Graptemys geographica*) were equipped with tracking tags emitting radio signals and found to range up and down the Lamoille River in Vermont and into Lake Champlain over at least a dozen kilometers. In the autumn as water temperatures drop quickly from 22°C in August to 11°C in September, and 2°C in November, the turtles congregate in one assembly about three kilometers up from the mouth of the river. This communal map turtle hibernaculum (which also includes softshell turtles, *Apalone spinifera*) was investigated by the biologists using scuba gear. They saw turtles pile on top of each other in a deep depression where there is negligible current. The turtles stay at the same site from November to the end of March. After the ice melts and when water temperatures warm up from 0.1° to 12°C, the turtles again leave and return to Lake Champlain for the summer (Graham et al. 2000).

After ice covers the Lamoille in December, the turtles are unable to come up for air for about five months. Do they experience stress of submersion? The biologists studying these turtles (Crocker et al. 2000) returned monthly to the communal hibernation site throughout the winter. Using a chain saw, Carlos Crocker (from balmy Alabama) cut a hole through the ice and then dove down and retrieved turtles and gathered environmental data such as water temperature and oxygen tensions. He sampled the turtles' blood to measure acidity, lactate, and oxygen and carbon dioxide concentrations. The conclusion drawn from the data was that these large thick-shelled turtles remain essentially aerobic (oxygen-breathing) all winter long despite their inability to breathe with their lungs. They avoid the progressive acidosis that results from anaerobic metabolism, suffering no apparent diving stress because their low metabolic needs for oxygen are met despite inability to take a breath of air into the lungs for months. Their oxygen needs are low due both to their physical lethargy and their low body temperature that reduces resting metabolism. How they accomplish any oxygen uptake at all is not clear. However, the hibernating turtles rest with their heads and legs fully extended on the river bottom and may thus be exposing as much skin as possible to take up dissolved oxygen from the water.

Our best understanding of the hibernation dive physiology of turtles comes from the painted turtle, *Chrysemys picta* (Ultsch et al. 1999). This species, like other water-inhabiting species studied, also shows no apparent diving stress under simulated hibernation dives in the laboratory at 3°C in normal, i.e., unaerated water. That is, they show relatively little rise in lactic acid and also no change in blood glucose. These results indicate that gas exchange through the skin is sufficient in these turtles as well, at least if they lay on the pond bottom at near 3°C. However, these turtles normally hibernate by burying themselves in the mud, which is nearly devoid of oxygen so that they apparently even deprive themselves of breathing through the skin.

In order to find out how the turtles respond to oxygen lack, the researchers (Ultsch et al. 1999) brought them into the laboratory and sealed them into water bubbled with nitrogen gas to drive off all the dissolved oxygen. The almost totally oxygen-deprived turtles then survived for "only" about four months at 3°C. Their blood lactate increased steadily throughout the whole time of immersion. Blood pH declined from slightly basic 8.0 to near-lethal levels of 7.1. The acidification of the blood (to near-neutral pH) was compensated for in part by increases in concentrations of positive ions (magnesium, calcium, and potassium) that buffer the acidity.

Southern populations of these turtles reached near-lethal blood pH in only thirty days, while western ones required four to five times longer to reach the same lethal levels. Eastern populations were intermediate. Thus, hibernation dives are different from ordinary dives, and different populations of these turtles have adapted to withstand the specific magnitudes of the stresses that they encounter in the wild during their hibernations.

As already mentioned, map turtles hibernate in depressions of the river bottom where warm flowing water in the spring from the river drainage could reliably signal the arrival of spring. In contrast, painted and snapping turtles inhabiting small ponds hibernate under mud in shallow stagnant pools close to shore (Ultsch and Lee 1983) even though it is stressful for them in that nearly oxygen-free environment. Ultsch and co-workers (1985) suggest these latter turtles choose the *shallow* water of stagnant ponds to hibernate in because of some as yet unknown advantage. Perhaps, because such water would heat up rapidly in the spring, it could provide the turtles their emergence cue and thus reduce the length of hibernation. Turtles have a relatively short season in the north to recoup energy

Young painted turtle.

losses, mate, grow, and lay and mature eggs. However, hibernation in shallow water that might promote getting an early start in the spring has a large drawback; predators such as raccoons could reach them there. Turtles therefore probably choose the physiologically more stressful burial in the anoxic mud because that behavior would reduce predation, especially in fall and early spring when the turtles are sluggish.

The ability of a turtle to buffer its blood with potassium and calcium ions, to reduce the acidity of lactic acid, contributes greatly to its winter survival under the ice. However, somehow this solution doesn't quite capture the whole reality of an amazing, astounding feat. It does not account for a turtle's just plain toughness and tenacity.

Turtles, both painted and snapping, often get run over by cars on the country road where I live as they travel to and from their nest sites. One warm June day when I stopped to pick up what I thought was a dead washtub-sized roadkill snapping turtle, perhaps one I'd met earlier in happier circumstances, I was left to ponder what life or death might be to a turtle. A dozen or so Ping-Pong-ball-sized round eggs were strewn all around this smashed turtle. As I touched its tail, the animal retracted its legs. Thinking the badly smashed turtle might perhaps still be alive, although I knew it could never recover, I wanted to put it out of its misery quickly. I maneuvered my pickup truck to run it over squarely. Another car came by just then and the driver, quite understandably, stared at me angrily. But the good and difficult deed was soon done nevertheless, and I dropped the turtle off with my ravens after chopping off its head (since the body still twitched). To my great surprise the birds had still not fed from it by the next day. As I pulled once again on the tail of the long-since-headless turtle, her legs contracted into the shattered remains of shell, as they must have if the ravens had pecked it.

What is death to a turtle? what is being alive? For six months it stays under ice water, buried in mud, where all breathing, movement, and presumably almost all heart activity stops. In spring it comes up, warms up, takes a few breaths, and resumes life where it had left off. It has done so for the perhaps 200 million years or so that its kind have prospered with little change. After the nineteen-mile-diameter asteriod struck the Yucatan Peninsula in Central America 64 million years ago and raised a dust cloud that caused the "global winter" that killed off the dinosaurs, they continued to live on as superbly successful and diverse animals to the present time. Only now, subjected to ecological effects from humans, are some populations endangered. Otherwise, they are still so well designed that they require little change. Maybe they survived that fateful global winter after the asteroid struck earth much like they now routinely survive a northern winter, by simplicity. They reduce their energy expenditure to extend their oxygen and energy reserves.

The snapping and painted turtles that come in June to lay their eggs in the sun-warmed sand in our driveway would not be here were it not for the beavers, whose dam meanders across a valley between wooded hills. The beavers predate all of us humans in this landscape, having been here for thousands of years, except when they were once temporarily driven out by trappers due to a fashion in hats in Europe. Their dam remained. Parts of it are ancient. It has likely been broken and torn out thousands of times, but it will always be repaired or rebuilt. This dam holds a shallow pond of several acres, which is where the turtles come from in June, and also where they return in the winter to be safely under the ice. Without beavers this would be unbroken forest. There would be no painted or snapping turtles, no bullfrogs, green frogs, mallards, Canada geese, dragonflies, giant predacious water beetles, snipe, Virginia rails, willow flycatchers, yellow warblers, red-winged blackbirds, sunfish, minnows, catfish, kingfishers, great blue herons, mink, or muskrats.

It is a rare day that I do not pause at this beaver bog to soak up the marvels. I record what I see to keep it for later. The following is a journal entry on the day before the final freeze-up.

9 Dec. 2001

It was 20 F yesterday morning under clear blue skies, but by afternoon high clouds were drifting in. Very promising—snow is surely on the way.

I wake up in the night, look out, and see white ground. I can not sleep any more as I anticipate a dawn with the first real, powdery snow, after the slush we've had so far. I get up, grab a cup of coffee, and head down the driveway to the beaver pond at 6 am. Since nights now last almost 15 hours, it is still night when I get there.

Fine feathery snow crystals drift down. There is not a breath of moving air. The sharp clean smell of this new snow prickles my senses and excites. Within a minute I stand at the edge of the pond feeling peace, and just barely hearing the tinkling of snow crystals falling on my jacket. They amplify the stillness.

The freeze-up is late this year. The pond was only covered yesterday with its first thin sheet of ice. Before that its surface reflected the dark shadows of the surrounding pines, now light from the moon that is barely visible through thinning clouds illuminates a white expanse except for two dark bare patches of water. There the snow is wetted into a layer of gray slush. One patch is around the beaver lodge at the opposite side of the pond near the dam, and the other patch is by the mound of decaying cattails where the geese nested in the spring. I can just barely make out two black lumps along the edge of this slush-patch. They look like stumps, but I don't remember ever seeing any stumps near there. They would not attract my attention, except that they have no white topping of fresh snow on them. There is snow on every blade of cattail leaf, and on every twig of alder and arrowwood bushes around me.

Nothing has changed twenty minutes later. It is still dark, and I've not heard a peep of a bird, nor seen a wiggle of the two black shapes although I imagined that one moved slightly. But in the dark, under gently falling snow, one can all-too-easily imagine all sorts of things, wonder about them, and come to absurd conclusions based on unrestrained imagining. After another 20 minutes, with the first lightening on the eastern horizon, come the first calls of goldfinches. In the daytime I recently had seen a flock of about 80 feeding on white birch seeds. The birds stayed closely bunched, and like one organism they synchronously and erratically flew off the tree, circled, and alighted again at the same tree. They hung like Christmas ornaments from the twigs and twirled around them to extract the seeds out of the fruiting cones. Showers of cone bracts drifted down and peppered the snow on the ground. The goldfinches are now, in the dawn, flying in search of another fruited birch, to begin their daily work of fattening up to fuel them through the next night.

There is a sudden whirr of wings from a heavy bird behind me. It is probably a grouse coming out of cover in the thick pines to feed on buds up in the bare branches of a poplar or birch. The snow is not yet deep enough to tunnel into to hide and stay warm.

A movement catches my eye: A long black shape comes

loping onto the pond from the edge of the cattail patch. A mink. After coming out onto the pond, the water weasel lopes over to the larger of the two muskrat lodges, briefly examines it, and continues on to the next one nearby, the smaller one.

I hear a musical twitter of a flock of snow buntings who are flying by, high overhead. They've come from north of Hudson Bay. They are the snow birds, and winter is not far behind. Two crows fly by, cawing loudly. The mink pauses a few seconds along-side the muskrat lodge, then runs on across the pond, past the two still immobile dark shapes along the slush hole, and then on into the cattails on the other side.

I stand as still as before. I'm mesmerized. It is getting light now. The snowflakes continue their soothing rustle on my jacket. A raven croaks in the distance, from where it comes every morning at dawn. Tree sparrows, migrants from the Arctic tundra, stopped in the bog a few days ago to fatten up on seeds. Their sweet-sounding sing-song notes resound back and forth, as they stay hidden close to the ground under the alders at the pond edge. They'll be gone in a few days. Suddenly I hear a tinkling-rustling sound right next to me. I look down and see a blade of sedge wiggle. A tiny load of snow slides off. A flash of movement. A moving black dot. It's the eye of an immaculately white weasel. The weasel disappears, reappears somewhere else as if by magic from under snow-covered grass. I see now a fresh mouse track. Probably deer mouse, given the tail-drag and the long stride. Flushed prey? The weasel dashes from one clump of grass to another, crossing the mouse's trail. The weasel stands up, extending its slender six inch body toward a tiny rustle, looks in that direction, then dashes off. In seconds it is back, standing tall and looking at me. Fearless, focused, improbably alert, and powered with unbounded restless energy, it soon again disappears from sight. (Following its track, I later found where it had dragged something, leaving drops of blood on the snow.)

I'm becoming ever-more curious about those two immobile black lumps on the ice. Are they the heads of otters peering up out of the water? Two miniature hunched-up beavers? Maybe muskrats.

It is light, finally, and one of the lumps slides into the slushy water. Muskrat it is. Its companion stays put. Within about one minute the diver is back, sculls briefly around the slush pool, and hauls out next to its companion. It sits up and preens its fur. The other looks around, and then also slips down for a dip. Their lodge looking like a miniature beaver lodge, is within a few seconds' swim under the ice, but the rats stay outside in the open on the ice. (I found them here off and on for the rest of the day whenever I checked.)

By the next day the pond was totally frozen over. The entrance to the rat's mud and cattail castle, of soon-to-be-rock-solid cattail-reinforced ice, was under water. It had been the rats' last chance to see daylight. Now they would be sealed in for almost half the year. Only the warm spring sun melting the ice will eventually release them. I suspect they have no notion of what lies ahead in the months to come—and for that matter, neither do I.

Usually when I gaze over the pond in the weeks and months after the ice has sealed it in, I see only the reminders of the pulsating life it supports and harbors. I see the sedges along the edge that in late April or May shoot up like sharp green lances. In winter they are bent into mounds and weighted down with snow. The cattails that hide the deep nests of red-wings cradling light blue eggs with purple squiggles are limp and lifeless. The nests of cedar waxwings, catbirds, goldfinches, and kingbirds are long abandoned and exposed on the leafless arrowwood bushes. They will soon fall to the ground and be reclaimed by the soil. However, three conspicuous structures rising from the water are ready for life. They are the shelters of water rodents, ready for occupancy. They were made for overwintering (see Chapter 5, "Nests and Dens").

Water is essential for the beaver's winter food supply. In the fall, beavers get busy, working, well, like beavers are supposed to. The whole family pitches in and fells trees in the nearby forest. Some of the mature poplars they have felled near my house measure up to 53 inches in circumference, but thankfully the animals preferentially harvest young, fast-growing trees that are dragged away whole.

Trees are limbed and the limbs dragged into the water and then floated out to near the lodge. By freeze-up the beavers have accumulated a brush pile of hundreds of pounds next to their lodge. It presses into the pond bottom and the top of the pile sticks out of the water. Foot-thick, concrete-hard walls keep the occupants safe and warm in the deepest cold. After freeze-up, whenever a beaver needs to feed, it must exit its lodge underwater and swim out to the food cache to bring back sticks to feed from. During such dives, as in other divers, the beavers' heart rate drops and energy expenditure is reduced to prolong diving duration (to about fifteen minutes).

Cross section of a new dam and lodge with family quarters.

6.5'

Winter food cache.

Clearcut

3'

The food cache, though large in size, is generally short in useable calories. Like many other herbivores, including termites and cockroaches, beavers compensate by harnessing cellulose-digesting bacteria in the gut. And then they recycle their rich cellulose diet once more, by eating their fresh feces. Despite all this, they also lay up body fat in the fall. Finally, adults (but not growing kits) save energy in the winter by tolerating greater amplitude of body temperature fluctuations, and reducing mean body temperature by about 1°C (Smith et al. 1991; Smith, Drummer, and Peterson 1994).

After the beaver family is imprisoned under the ice and in the lodge, the only source of air they have access to is through a small vent-hole at the top, which lets in air through a latticework of thick sticks.

The lodge so prominently visible to me near the opposite shore, beyond the hunkering muskrats, would soon be frozen in. What might it be like to spend months huddled in near-absolute darkness, except for a perhaps once-daily dive to feed in the snow-covered, inky, ice-cold water? I surmised that, having adapted to these conditions, beavers could not be too unhappy in them. They are like muskrats, probably placid creatures, and likely not as claustrophobic as I.

Beavers and muskrats are small-eyed primarily nocturnal animals, and their activity is governed by a circadian schedule. We normally start our daily activity when darkness changes to light. But a beaver or muskrat becomes active by leaving its already-dark lodge in the dark of the night or just before dark. How does it know that it is time to get up and going? The proximate answer is that as in us and flying squirrels their circadian rhythms alert them when it is the right time. However, like cheap watches, biological clocks would eventually get out of phase with the day/night cycle. To be useful, they have to be reset periodically by some cue, to synchronize with the external environment. We set our biological clocks using the lights-on transition as a reference signal. Without such a signal, our internal rhythm would gradually bring us out of phase with the external environment, or we would be unable to readjust when we pass into another time zone. Beavers in the winter can apparently lose contact with external light-dark signals, and their activity rhythm, which is slightly longer than twenty-four hours, starts to free-run. As in DeCoursey's experiments with flying squirrels in constant darkness, each animal gets more out of synch with every passing day (Bovet and Oertli 1974). That is, the beavers probably *experience* constant night. Of course this makes no practical difference to them in the perpetually dark, safe world under the ice. A schedule is then irrelevant.

Muskrats presumably experience a constantly dark winter world similar to that of beavers, and they have evolved to solve the same problems of energy shortage and keeping warm. But muskrats don't build dams. The two I had watched at dawn depend on the water provided by beavers. They build a house specifically for winter. I had seen two in front of me near a patch of cattails, and the one closest to me, like most others, is a conical, two-foot-high pile of dried cattail leaves scraped into a heap and patched together with mud dredged up from the bottom of the shallow water where it is built.

The muskrat's house is a partial solution to extreme cold as studied in detail by Robert A. MacArthur from the Zoology Department of the University of Manitoba. Muskrats also huddle, but weighing only about 2 pounds to the beaver's 40 to 60 pounds, they lose heat more easily and their need to huddle is greater. That need for heat is met by becoming more tolerant for fellow muskrats. Even nonkin may gather together in a lodge and thereby gain several advantages. They warm the lodge, huddle (Bazin and MacArthur 1992), groom each other, and apparently stimulate each other to go out and forage (MacArthur, Humphries, and Jeske 1997). Beyond that, the muskrats rely on physiology to a greater extent than beavers do, making up for what they don't solve by behavior.

Like beavers, muskrats are nonhibernators that maintain a high body temperature and thus need continuous fuel for their high metabolism. Unlike northern beavers, however, they don't collect a cache of food prior to winter (nor do southern beavers). They are therefore forced to continue foraging. They feed on plants on the pond bottom, but being locked in under the ice presents problems for a warm-blooded air-breather. Unlike the torpid turtles, they need to inhale oxygen, and a lot of it, because swimming is hard exercise. Their only source of oxygen may be in the lodge, where its concentration can be low and that of carbon dioxide high, due to the respiratory needs of fellow lodge occupants and slow rate of gas diffusion through the solidly frozen lodge walls.

Muskrats have solved their problem of access to oxygen like other mammals that dive for a living. But they do it better specifically in the winter (MacArthur 1984b). In the winter muskrats carry more oxygen in the blood by increasing the number of red blood cells with their oxygen-binding molecules, hemoglobin. They also have stores of oxygen in the muscles, where it is held by a special protein, myoglobin. Myoglobin is the oxygen-binding protein that colors meat red. With 42 percent more oxygen stores in the body in winter than in summer (MacArthur 1984b; 1992a), muskrats gain more underwater foraging time and/or foraging range. Additional foraging range is achieved by making feeding shelters in the fall that look like miniature lodges, reaching about a foot above the water surface. I suspected that the second, smaller, muskrat lodge that I saw was one of these. Muskrats may also make "push-ups" of vegetation later on, on the ice where they can find cracks (MacArthur 1979). Both kinds of shelters are built within swimming range of the main lodge and serve like the breathing holes that arctic seals maintain. Here they can come up for a breath of air or to feed on roots brought up from the bottom. Additionally, they may exhale air bubbles that get trapped under the ice, and air from these bubbles can later be used to extend dive durations (MacArthur 1992b).

Limits of underwater foraging in winter also depend on temperature. Beavers and muskrats are of a select group of animals that can swim in ice water because of their extraordinary fur, which keeps their skin dry by trapping a layer of air next to their skin. This insulative air layer, which solves some of the problem, is the result of a luxuriantly thin and fuzzy underfur and highly specialized thicker "guard hair" that is long and glossy and protects the underfur (felted into hats, it almost spelled the beaver's total demise).

The beaver's feet and tail stay cold and are not furred. Like the kinglet's legs, the beaver's and muskrat's feet and tails have a specialized blood circulatory anatomy that helps to prevent body heat from escaping from these body parts. The principle is that if it is costly and difficult and not necessary to keep the extremities warm, then it's better to try to economize and keep them cold. Nevertheless, despite these adaptations of keeping the body hot, rapid heat loss is still inevitable during prolonged dives into ice water.

When a muskrat leaves its lodge and submerges itself under the ice, it not only diminishes its oxygen supply, it also immediately starts to cool. As it forages, it may periodically come up to breathe at a push-up or a feeding platform, where it can replenish its oxygen supply, but it can still be losing heat at a high rate. Most voluntary dives by muskrats last less than forty seconds, although the rats can store enough oxygen to stay down for several minutes. An active rat does not, and presumably cannot, allow its body temperature of 37°C (similar to ours) to drop more than 2°C (MacArthur 1979), and the rat's solution, like that for lack of oxygen, is to store up a surplus of heat. Just prior to diving into ice water in winter (in contrast to summer), a muskrat increases its body temperature on average by 1.2°C (MacArthur 1979). Then, after returning to the lodge, the rat shivers and expends energy at a high rate to heat itself back up to 37°C (MacArthur 1984b).

Although the rats' physiological response suggests that their dip into cold water is anticipated, do they really plan ahead? Perhaps, but if so, then the white-faced hornets I have mentioned previously in reference to their insulated papier-mâché nests may do so as well. Some years ago, I took on the brave, or foolish, task of measuring hornets' body temperatures, grabbing and stabbing them with an electronic thermometer as they left their nests. I learned by my experiments that prior to leaving their nests at low air temperatures (2°C), the hornets shiver to heat themselves up to 39°C. On the other hand, when they leave on a warm day, at 22°C, they warm themselves up to a body temperature of only 35°C. (When *attacking*, they heat all the way up to 41°C.) That is, they appear to be as clever as a muskrat, or vice versa. I suspect, however, that body temperature is much too important and constant a concern for any animal to be able to rely on mere cleverness to consistently produce the correct responses.

In the fall *after the birds have left, I often hear*
birdlike chirps coming from the woods, especially
after it warms up for a few days. I've tried numerous times to sneak up carefully and identify a
caller, but so far I have discovered little more than
that it was hiding on or near the ground. Whenever
I've come close enough to almost step on the
sound source, it went silent and I saw nothing.

By chance I found a passage in the writings of
the famous nineteenth-century American
naturalist-writer John Burroughs, who succeeded
in tracing apparently similar sounds. Walking in
the woods in his native New York on the last day of
December 1884 when it was so unusually warm
that bees were flying outside their hives, he paused
in the shade of a hemlock tree where he heard a
froglike sound from beneath the wet leaves on the
ground in front of him. Determining the exact spot

from where the sound came, he lifted up the leaves and found a wood frog. "This, then" Burroughs concluded "was its hibernaculum—[where] it was prepared to pass the winter, with only a coverlid of wet matted leaves between it and zero weather." (We now know a great deal about how amphibians survive the winter, but we still don't know why some of the ones that have loud mating choruses only in the spring after the ice goes out, may also sometimes call individually in the fall.)

Amphibians, especially toads, were known to dig into the ground to escape frost. Gardeners, your narrator included, often find them when turning soil in the fall. Some toads go down deep, as John R. Tester and Walter J. Breckenridge, two biologists from the Museum of Natural History at the University of Minnesota, Minneapolis, learned while studying the Manitoba toad (*Bufo hemiophrys*) in the Waubun Prairie in northwestern Minnesota. The area is known for its extremely cold winters. The toads are associated with water in the numerous potholes, and they travel 75 to 115 feet from water to burrow and hibernate in gopher mounds. In three years, the two researchers collected 7,483 toads from eight mounds, tagged them with radioactive chips (100 microcuries of tantalum-182), and then periodically traced their whereabouts from above ground with a portable scintillation counter. Their study revealed that the toads dig even while they are in hibernation, going ever deeper throughout the winter. They dig down to four feet or deeper, staying just barely ahead of the ever-deepening frost line. They stop digging and stay at temperatures 1° to 2°C above freezing, which is then also their body temperature.

Burroughs reasoned that surely his wood frog would have, like a toad, buried itself into the ground if it anticipated a severe winter. However, since the frog had remained at the ground surface that would soon freeze solid, Burroughs thought that surely a mild winter was in the offing. Instead, a severe one followed: The ice on the nearby Hudson River was nearly two feet thick, and it was bitter cold even in March when Burroughs went back to reinvestigate the frog in its hibernaculum under the leaves.

The matted leaves were then frozen hard. Burroughs lifted them and found the frog "as fresh and unscathed as in the fall" even as the

Wood frog
(Rana sylatica).

ground beneath it "was still frozen like a rock." He wrote: "This incident convinced me of two things, namely, that frogs know no more about the coming weather than we do, and that they do not retreat as deep into the ground to pass the winter as has been supposed." Undoubtedly Burroughs would have been even more convinced about how little frogs know about the coming winter if he had found one frozen solid. He wrote that the frog he found on the rock-hard frozen ground "winked and bowed its head" when he touched it.

Fast forward now almost a hundred years later, to William D. Schmid, a comparative physiologist at the University of Minnesota in Minneapolis. Schmid had previously studied the tolerances of frogs to dehydration, and like Burroughs he also made a serendipitous discovery of a wood frog shallowly hidden under leaves in the winter. But Schmid's observation would soon revolutionize accepted ideas, largely because he made a second observation: While handling the frog he noticed that it did *not* do what handled frogs usually do; which is to wink. Instead, it appeared to be frozen solid.

Having previously learned that different frog species have appropriate adaptations to survive in their unique habitats, Schmid doubted that the frog he found had made a lethal mistake in not burying itself deeply enough or in choosing the wrong spot to spend the winter. He therefore followed up his hunch that the frozen frog might still be alive, and a now-classic study of cold-weather survival in frogs followed. He published it in the prestigious journal of *Science* under the unassuming title "Survival of Frogs in Low Temperatures." As far as I know, this is the first documentation of freezing-tolerance as a winter survival strategy of any vertebrate animals. Perhaps Burroughs had not believed his senses when he found his wood frog to "wink" underneath frozen foliage and on top of rock-hard frozen ground; given the conditions he described, the frog that he found *should* also have been frozen solid. To most animals, freezing means certain death.

Kenneth B. and Janet M. Storey at Carlton University in Ottawa took up the banner of frost-tolerance and explored the physiological underpinnings of freezeing-tolerance in frogs. They confirmed and extended what Schmid had discovered, and we now know that four common North American hibernating frogs—the wood frog, gray tree frog, spring peeper, and chorus frog—all tolerate being frozen. In freezing-tolerant frogs there is extensive ice formation in the body cavity and in the spaces between the cells (up to 65 percent of the total body water in the wood frog may be ice), but in frogs that survive there is no ice crystal formation within the cells themselves. Ice crystals are normally lethal when they form within the cells because they cut like knives, slashing membranes, puncturing cell organelles and breaking cells. The frog's lethal low temperature limits go only as low as -8°C because at lower temperatures than that ice does form within cells. But lower temperatures are seldom encountered in wood frogs' hibernacula under leaves and snow. (Where spring peepers, chorus frogs, and gray tree frogs hibernate is not well known. I've often heard the peepers piping in the woods in the fall, so they presumably hibernate somewhere in the woods. One colleague told me of finding a tree frog under loose bark in the winter, and another had one hibernate on a houseplant she brought inside in the fall.)

All of the four above-mentioned frogs that hibernate on land can tolerate about half of their body water turning to ice, but that feat is not possible without the aid of chemistry that addresses two main problems. First, one chemical (primarily the alcohol glycerol) protects membranes when freezing does occur, and second, the other (primarily glucose) is mostly but not exclusively involved in an osmotic response that restricts ice-formation to outside the cells. However, unlike in the cold-hardening of many insects that similarly use alcohols and sugars to perfuse their tissues prior to winter, frogs do not accumulate these chemicals in the fall in anticipation of freezing. Instead, frogs *wait* for the ice to form, and in one day they change themselves to become frost-tolerant. The ability to tolerate freezing is acquired in a modification of the adrenaline-mediated fright response of other vertebrate animals, such as in ourselves.

All vertebrate animals have a fight-or-flight response in which the sensory input of a threat, say a charging lion, causes the hormone adrenaline to be released from the adrenal glands. Adrenaline has wide-ranging effects, but the net effect is to prepare the body to meet the challenge. Heart rate increases, blood glucose concentrations rise, and blood flow is redistributed to the muscles. This adrenaline response has been modified in wood frogs to meet the freezing challenge that is lethal to aquatic frogs, and of course to all of us.

When the first ice crystals begin to form on or in the skin of a wood frog, it sets off an alarm reaction. Skin receptors relay the message of freezing to the central nervous system (CNS), and the CNS activates the adrenal medulla to release adrenaline into the bloodstream. When the adrenaline circulates to the liver, it there activates the enzymes that convert the liver's stores of glycogen to glucose. As a result, the frog responds with a quick rise in blood glucose. In the wood frog, this response is massive and before the ice reaches the cells they become packed with glucose that acts as an antifreeze.

Precisely the opposite occurs outside, between the cells, where special proteins act as ice-nucleating agents to *promote* ice crystal formation in areas of dilute fluid. As a result, pockets of concentrated fluid are created, and these act to osmotically withdraw water from the cells, making them even more resistant to ice formation. In about fifteen hours, the frog is frozen solid except for the insides of its cells. Its heart stops. No more blood flows. It no longer breathes. By most definitions, it is dead. But it is prepared to again revive at a later date.

Glucose is the normal vertebrate blood sugar that is used by the cells for energy. In the healthy, active animal, blood glucose concentrations are normally precisely regulated by two hormones, insulin and glucagon. When we don't have enough glucose in the blood we become unconscious, and when we have too much we suffer numerous short- and long-term consequences. We normally regulate our blood glucose at near 90 micrograms per 100 milliliters of blood, although during stress these levels rise slightly. When levels of blood glucose reach and are chronically maintained near 200–300 mg/ml blood, we're diagnosed as ill with a disease called diabetes mellitus. Diabetes can now be treated by administering artificial insulin, the hormone that the pancreas is not producing in sufficient amounts in one version of the disease. (In Type I diabetes the pancreas is destroyed in an autoimmune response. In Type II there are no or few hormone receptors.)

In frogs that have felt ice crystals forming on as little as a toe, the massive blood glucose levels (to 4,500 mg/ml of blood that are released in a seeming hyper-alarm response), would be high enough to send us into a coma and death several times over. But to the frogs, which survive it, in part because they are then at near 0°C, and metabolically relatively inert, it is their ticket to survival, to tolerate freezing.

Soon after the frog's heart and breathing stop, its tissues would become starved of oxygen if metabolism continued. However, at high concentrations the glucose acts as antifreeze, a mechanical protectant from ice crystals, and an agent to help draw water from the cells. It also reduces the frog's already very low aerobic metabolism and thus acts as a metabolic depressant to conserve the cells' limited energy reserves for the winter. The glucose that enters the cells also becomes a substrate for anaerobic metabolism when the body can no longer supply oxygen.

Frozen bodies that can revive upon thawing out have long been a pipe dream of cryobiologists. The frogs that hibernate in the forest floor do it routinely, coming out of their frozen state at the first flush of spring when it is time to mate. They, like the hibernating bears I'll discuss later, are biological marvels that challenge the limits of our beliefs of what seems possible.

It seems astounding to us that some frogs can survive months being frozen, or that a bird as small as a kinglet can stay warm and survive even one winter night, much less a whole northern winter. But why, really, are we surprised? I think it's because we compare them to ourselves. We feel uncomfortable when we chill only a degree or so and we can't imagine how a tiny bird keeps warm in a blizzard. Yet for every kinglet that we find in the winter woods hundreds of thousands of invertebrate animals exponentially tinier than a kinglet survive by doing what for us seems unimaginable. Even when we do know what they accomplish we still tend to withhold respect. Why? It's because most of them are insects. They are animals so different from us that it's as if they were from an alien world; we find it difficult to identify with their problems. Yet they face the same problems of cold, freezing, and energy balance that we or a kinglet deal with. They have evolved some of the same, and also different solutions, but with different constraints.

To an entomologist and anyone who aspires to be one, there are no life-forms on earth as diverse, varied, tough, and inventive as the insects. In their teeming millions of species, they own the world. We may not like many of them that compete with us for food, fiber, timber, or that suck our blood and spread our diseases, but we are obliged to acknowledge their tenacious success, and we may admire many of them for their stunning beauty. Within the animal world they have collectively pushed the limits of things possible, in terms of diversity, beauty, noxiousness, social organization, architecture, powers of flight, sensory capabilities, and ability to survive extremes of climate. And when I contemplate these organisms that are much more ancient than us, and that will long survive us, I wonder about the "secret" of their success and then I am forced to confront how differently a physical scientist and a life scientist sees the world.

The physical scientist tries to understand the world according to mathematical precision by reducing its composition and functioning to a very few "laws of nature." Such laws, in analogy with our own civil laws, are as if handed down from some higher authority who then enforces them by virtue of that authority because a law is, by definition, general. It applies uniformly. There are no exceptions, or no exceptions are allowed.

Insects' success is derived from exploiting individual specificity. No one way is best. Insects achieve their success *through* their diversity, where each individual case is special within the generalizations. Each species is adapted ever more specifically into a specialist niche, catering to specific individual needs. An ever-greater narrowing down to the specific has resulted in miniaturization, and ever-greater diversity. Insects exhibit an exhilaration and a celebration of the exceptions, where anything goes that can. There are few boundaries, because there has been no enforcement or encapsulation by or in laws. *That* is why they are so successful, and I suspect that it would be difficult to find an entomologist who is also a theist, who believes that there is a force or a power that hands down rules because "he" deems them good. No entomologist could fathom why fleas, mosquitoes, tsetse flies, migratory locusts, and dozens of other insects would have been deliberately created and let loose to cause indiscriminate and unimaginable agony to millions of totally innocent human children and adults over all the ages of humanity.

And it is this indiscriminate capacity to individually do what is best for them that has allowed insects to advance far into the winter world, one species at a time, to explore and exploit the many possibilities.

My fascination with insects started before I was ten years old, thanks to my father, who took me with him on winter adventures to look for hibernating ichneumon wasps. The white grub-like larvae of these insects feed inside (and eventually kill) caterpillars after the wasps inject their eggs through the caterpillar's skin with their sharp hollow ovipositor (that also serves as the stinger in bees). Papa collected inchneumon wasps as passionately when he was eighty years old as when he was twenty. At that time he had only barely scratched the surface of their

Promethia moth caterpillar and ichneumanid parasite.

astounding abundance and diversity, even in Maine, and many of these exquisitely elegant creatures that keep other insect populations from exploding, still remained to be discovered.

The *males* of these and almost all Hymenoptera, the family of ants, bees, and wasps to which ichneumonids belong, don't solve overwintering. They simply *all* die by fall, by which time they are

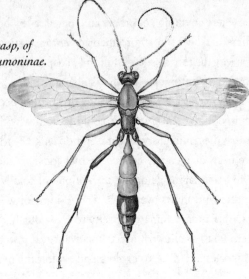

*An ichneuman wasp, of
subfamily ichneumoninae.*

superfluous since the overwintering females are by then insemi-
nated. Only the females seek overwintering sites in places where
there is moisture so that they do no desiccate. Temperatures must
also be low enough to dampen the metabolic fires and thus conserve
the limited stores of energy that they lay up in body fat. As with most
insects, however, temperatures cannot be so low as to cause freezing-
injury. Typically when unadapted soft tissue thaws out after freezing
solid, it turns into a brown mush. A quick routine way to kill
summer-active insects (such as specimens for one's collection, or the
pests that are eating one's collection) is by putting them into the
freezer compartment of the refrigerator at about -10°C. This is a very
modest temperature, relative to those that some species routinely
survive in the wild, in northern winters.

We found the ichneumon wasp females at three kinds of overwintering sites. We had the most success finding a variety of species
in the decaying wood of old moss-covered tree stumps. Other
species turned up only under moss, and still others were revealed
when we pried apart sedge or grass hummocks. With hatchet and

knives, we sometimes rendered a dozen or more stumps to mulch and found not a single insect. After all this work and anticipation it was exciting to suddenly find even one of these beautifully colored little wasps tucked comatose under moss or in decaying wood. Occasionally we hit a jackpot and found a dozen or more of them in a single stump along with an assortment of spiders, centipedes, millipedes, ground beetles, beetle larvae, and occasionally even an overwintering hornet or bumblebee queen. The insects we found could crawl only very slowly in the cold, but they quickly became lively in the warmth of one's hand.

Despite our frequent and long-standing searches in the winter woods of Maine, there were many species in my father's collection that we never found in hibernation. Perhaps they have specialized hibernation sites that we didn't know about. We found some species in hibernation that we never saw during the summer, so that our winter searches served as a way to enlarge our survey of the diversity of species.

Virtually nothing was known then of the physiological and behavioral mechanisms insects use to survive winter. But thanks to numerous researchers and a huge amount of work in recent decades, we now know that even in any one locality the different insect species exhibit a diversity of winter survival tactics. As previously indicated in the overwintering moth larvae that the kinglets feed on, a few species have amazingly evolved to survive being frozen solid, sometimes at temperatures much lower than those that would kill freezing-tolerant frogs. Some seek special shelter. A tiny number migrate. Many species avoid freezing by physiological mechanisms that lower the freezing point of their tissues. No one strategy is best since each has envolved under a different set of constraints and opportunities. The widest variety of ways of surviving winter is exhibited by the Lepidoptera (moths and butterflies), and I turn to them for a comparative view to explore the range of possibilities. They are also the best studied so far.

Spring azure pupa and adult.

Mourning cloak butterfly.

From northern New England and all the way across Canada, Europe, and Alaska, one of those which stays and overwinters as an adult is the mourning cloak butterfly (*Nymphalis antiopa*). In the summer the massed crowds of their black-and-red spiny caterpillars commonly feed on willow. Two generations are commonly produced per year, and second-brood caterpillars that I've reared in the fall always pupate and produce their adults before winter. The mourning cloaks, like many other butterflies in their family, the Nymphalidae, overwinter only in the adult stage. Mourning cloaks hibernate in hollow trees and these long-lived (in excess of 10 months) butterflies feed on fermenting tree sap such as a sapsucker licks. I never fail to see them in early spring as they bask with open wings in sunshine usually long before all the snow has melted.

Probably my favorite early spring butterfly is the spring azure (*Celastrina ladon*). Those brilliant blue mites that, like winged jewels, flutter over the just-emerging brown earth after the long winter. Females mate on the first day after emergence, lay their eggs on the second, and rarely live beyond the third. The larvae are tended by ants, and the pupae then spend most of the summer and then the winter in diapause.

In all overwintering insects only one life stage is adapted to survive winter. In the eastern tent caterpillar moth (*Malacosomia americanum*) that stage is the egg. The moth lays her eggs all in one bunch in a ring around a cherry or apple twig where they are encased in a foam that hardens so that the eggs become solidly attached to the twig. The eggs are exposed to the winter's lowest air temperatures but what shields them from the cold is glycerol, an antifreeze chemical that used to be commonly poured into car radiators in the fall. In early spring, the glycerol is depleted from the eggs and the larvae hatch. The group of larvae emerging from any one egg mass then spin the silk web in a fork between some branches that constitutes the familiar "tent" from which their name is derived. After completing their growth in early June, the larvae leave their tents and wander about, searching for places to pupate. Their light yellowish cocoons are placed into crevices under bark, and are a common sight on the sides and corners of buildings. By late summer the moths emerge, lay their eggs in time for overwintering, and die. Tent caterpillars, like most caterpillars, are unable to survive the freezing temperatures of winter.

The tent caterpillars are not endearing to most people, but the black-and-rust-red banded woolly bear caterpillars come close. They are the larvae of the Isabella tiger moth (*Pyrrharctia isabella*) that is specialized to overwinter in the caterpillar/larval stage. The Isabella tiger moth (formerly *Isis isabella*) is a member of the family Arctiidae, in which most species are beautifully colored in striking patterns of bright pink, red, black, yellow, orange, pure white, and blue while in the adult, moth, stage. The Isabella moth, in contrast to most of its group, is plain colored with predominantly dirty-yellow forewings and with a pinkish-yellow hue to the hindwings. W. J. Holland in his classic *Moth Book* (originally published in 1903) writes: "Both the moth and the larva are common objects, with which every American schoolboy who has lived in the country is familiar; and unhappy is the boy who has not at some time or other in his life made the country his home."

I'm happy to report from the countryside of Vermont and Maine that the "banded woolly bear" is still familiar to most people. According to local folklore, the width of the central reddish band reflects the severity of the coming winter. However, the caterpillars subtly change color over successive molts through the summer, becoming less black and more reddish as they age, i.e., as winter approaches. When touched, this caterpillar characteristically curls up into a defensive posture with stout bristles sticking out in all directions, much like the European hedgehog. In this species as in other Arctiids, only the caterpillars survive winter.

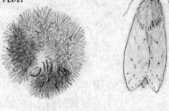

Woolly bear caterpillar in defensive and hibernating posture. Isabella moth at rest.

In the fall of 2001, I picked up three woolly bear caterpillars and I wondered how they, as well as three hatchling snapping turtles that I had retrieved as they were digging out of their nest in sandy soil, might bury themselves for winter. If I let these animals loose I could, of course, never hope to find them in winter. My question was simple: Will they bury themselves in moist soil, hide under the dead leaves, or stay on top of the leaves under the snow? To find out I filled a two-gallon plastic jar with soil, put leaf litter on the soil, and then buried the jar in the woods up to the level of the soil. Within a month frosts had hardened the top layers of the soil, and the jar was buried in snow.

Winter was not yet over by the end of February, but whatever the caterpillars (and turtles) were going to do to prepare for it, they should have done by then, so I shoveled off the snow and pulled the jar up to examine its contents. I found the three caterpillars almost immediately. They were unfrozen and curled up just under the leaf litter. The turtles were not in sight. Due to the snow insulation, the ground was, as usual, unfrozen, and I dumped the soil from the jar to sift through it. It was difficult to find the turtles. They were caked in mud with their heads and feet retracted into their shells and tails curled alongside. They looked like muddy pebbles. I distinguished them from pebbles at first only because they were compressible with my fingers.

To find out how these hibernators might act when warmed up, I brought them all, caterpillars and turtles, inside the house. After being washed with water at room temperature, the turtles "instantly" extended their limbs and heads, started rubbing their eyes and swimming in a bowl of water. The next day they started feeding on raw meat that I offered them. Their hibernation was over. Maybe they had simply been inactivated by the cold, after perhaps being induced by the cold to bury themselves.

The behavior of the turtles, as such, may not seem unique or surprising, except when seen in terms of the caterpillars' behavior. The three caterpillars remained on my desk in a jar in moist moss and green grass (one of their food plants) retrieved from under the snow in a field. I had expected them, like the turtles, to immediately start wandering and feeding on the grass when they warmed up. They didn't. They moved only enough to crawl under the moss and, even while experiencing the warmth of my study, to curl up again in the same hibernation posture that I had found them in. One month later the woolly bears had not moved. They seemed dead. Suddenly, in the last week of March, they all encased themselves in lightly spun cocoons (that incorporate the spiny hairs that they shed).

When I tested the frost-hardiness of the *pupae* by subjecting them to moderately low temperature (-14°C in the freezer compartment of our refrigerator) they froze solid and were dead.

It had previously been reported that cold-hardened woolly bear caterpillars remain unfrozen, even down to about -30°C, through a combination of supercooling and antifreeze. Low temperatures in the fall were reported to stimulate them to convert their glycogen stores into glycerol and sorbitol, and the amounts of these alcohols (up to 5 percent body weight) reduced the freezing point of their blood to about -10°C, and the rest—the prevention of ice formation of the whole animal down to -30°C—was presumably due to supercooling. Could this really be true for the New England population? I recalled having sometimes found woolly bear caterpillars in outdoor winter woodpiles in Vermont, and although I had no reason to study them, I did have the impression that they were occasionally hard frozen. But I didn't test if they were dead.

I wanted to put my woolly bear caterpillars to a test, and when I found two of them just out of hibernation (the following spring as I was writing this on Easter weekend) I put them (as previously the pupae) into a film vial and to -14°C in the freezer compartment of our refrigerator. Two hours later they were indeed quick-frozen into blocks of ice. They were solid. I could tap the table with them. When thawed out an hour later, they were alive and well!

This had been a severe test, since the caterpillars had already spontaneously aroused from hibernation and since freezing-survival (as I'll show later) requires *slow* freezing. Not believing my senses, I immediately repeated the experiment with the same two caterpillars. The result was the same: Woolly bear caterpillars (like the aforementioned geometrical caterpillars that the kinglets eat) do survive freezing—even multiple freezing—whereas the pupae don't. No wonder my caterpillars had waited so long to pupate after coming out of hibernation.

Sawfly cocoon (left)
on beaked hazel twig with
male catkin bud (right).

Cecropia moth cocoon on
red maple twig.

Some insects overwintering as
pupae in exposed cocoons.

Another woolly caterpillar, *Gynaephora groenlandica* (unrelated to the tiger moth), that lives in the High Arctic, has no chance to escape freezing solid. It is routinely subjected to the very much lower temperatures (to near -70°C) on the tundra where snow cover is thin and the ground is permafrosted. This species is one of the very few moths that has evolved to live within 83 degrees of the North Pole. During the short arctic summer the caterpillars briefly thaw out and feed. Because temperatures are low even then, they spend most of the year frozen solid. They only grow slightly in any one year before again freezing solid and their freezing-thawing cycle is repeated thirteen to fourteen years before they are finally ready to spin a cocoon on an exposed rock to catch direct sunlight for heating. They then molt, first into a pupa and then into adult moths that mate, lay eggs, and die a few days later.

The Gynaephora caterpillars living near the North Pole are surely exotic, and very few people get to see them (I was one of the lucky few who was invited by Jack Duman and Olga Kukal, two colleagues from the University of Notre Dame, to travel north to study them). However, there are also marvels at one's doorstep. At the farm in Maine, I collected nectar-sipping sphinx moths humming around the milkweeds by the barn. Later, my mothing took me to Los Angeles, where in the lab of George Bartholomew at the University of California, I tried to decipher if and how they regulated their body temperature. A couple of decades later I came full circle and returned east and discovered winter moths for the first time. These winter moths are not seen by the average person—only by those who go out into the woods at night with a bucket and a brush and paint trees with an ambrosiacal concoction of fermenting mashed fruit (apple, banana, or peach will do) spiked liberally with beer or some other alcoholic beverage. Each lepidopterist has his or her own special concoction that works best, which I suspect has as much to do with individual taste as with science.

Adults of the winter moths (*Cuculiinae*) don't just survive the winter in torpor. They *live* as adults in the winter world. There are numerous species of these moths belonging to the Noctuidae, or owlet moths, which is possibly the most species-rich group of Lepidoptera on earth, with thousands of species in temperate and tropical regions. The cuculiinae are distinguishable from other noctuids primarily by their so-called reversed life cycle. Whereas the vast majority of moths overwinter as pupae, the cuculiinids overwinter as adults and they also fly during the winter thaws when temperatures reach near or slightly above the freezing point of water. By mating and then maturing their eggs in the winter and laying them on the just-opening leaf buds in early spring, the adults are less likely to be eaten by bats, and the larvae also encounter less bird predation, since growth to the pupal stage can be finished before their predatory migrants return to reoccupy the northern woodlands.

Like sphinx moths and other noctuids, these winter moths have robustly built thick bodies powering short wings that require a high wing-beat frequency to support flight. In order to generate sufficient power for rapid wingbeats they must warm up their musculature to over 30°C. They do that by shivering. A shivering moth extends its antennae, raises its wings, and then you see wing vibrations as the upstroke and downstroke wing muscles are activated nearly simultaneously. A covering of thick insulating scales on the thorax acts, like fur or feathers, to approximately half their rate of heat loss. Further heat retention from the working thoracic flight muscles is enhanced by countercurrent mechanisms of the blood circulatory system that reduced heat loss into and from the abdomen. These moths are unique in their willingness and ability to start shivering when their muscle temperature is as low as 0°C. (Most others require 15° to 20°C higher temperatures.) Once shivering begins and the muscles begin to warm up, then shivering proceeds more vigorously, to produce even more heat until the suitably high muscle temperatures needed for flight are attained.

THE LEPIDOPTERA DEMONSTRATE diversity of adaptation to winter within a single group. However, a discussion of frost-hardiness must necessarily include a specific and very famous maggot, that of the goldenrod fly (*Eurosta solidaginis*). *Eurosta* is as necessary to an understanding of insect frost-tolerance as the fruit fly *Drosophila* is to genetics.

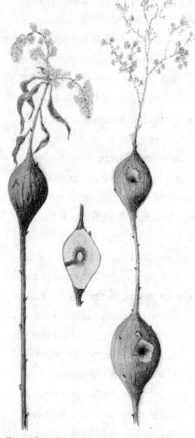

As implied by the name, the fly's life cycle is inextricably bound to the goldenrod's. The adult fly injects an egg into a young and rapidly growing goldenrod stem in the spring or early summer. Chemicals that are either injected with the egg or produced by the young larva then subvert the plant stem's normal growth, causing it to produce a thick tumorlike growth, called a gall. The gall has soft tissue on the inside, and is enclosed within a hard woody exterior. From within it the larva taps the plant's resources and uses them for its own growth.

In the goldenrod fly, as in the previously discussed Arctiid moths, only the larval stage is physiologically specialized to overwinter, and the insect's life cycle of one generation per year is adjusted to bring the

Goldenrod gall, showing cross section with gall fly larva (center) and two galls on one stalk, both excavated by downy woodpecker (right).

larval stage to winter. After passing through the winter the larvae pupate and then emerge as adult flies in time to parasitize tender goldenrod shoots. By late summer the goldenrod, the gall, and the larva have stopped growing and the larva then chews an escape tunnel from the center of the gall all the way to, but not through, the outermost edge. Retreating back into the center of the tough woody gall, it spends the winter there, in hibernation. That the larva makes the escape hatch when it does is essential, because the adult gall fly does not have chewing mouth-parts and it would otherwise remain entombed within the gall. Having prepared both an overwintering site and a means for the fly emerging in the spring to escape from it, the larva next prepares physiologically for meeting the winter cold. Northern populations of the fly have different responses than the southern.

Northern *Eurosta* larvae prepare for winter by producing both glycerol (an alcohol) and sorbitol (a sugar) in response to lowered temperature in the fall. The lower the temperature, the more glycerol and sorbitol they produce and the lower the freezing point of their blood. But their frost-hardiness doesn't end there. It involves far more complex and sometimes counterintuitive mechanisms, all acting in concert, that not all readers will likely want to follow. This includes, for example, the paradox that the northern larvae produce and release a protein into their blood that *promotes* freezing. In effect, the protein mimics ice crystals by providing nucleation sites for ice crystal formation. It thereby prevents the animal from achieving supercooling. By *preventing* supercooling, the protein causes the larvae to freeze earlier, already at *higher* temperatures than they would otherwise.

How can the apparent *promotion* of freezing aid in winter survival, even as the animals produce compounds with antifreeze properties? The answer is complex, and elegant. It relates to the fact that *very* low temperatures may be encountered in the northern larvae populations, and antifreeze alone would then be insufficient to guarantee absence of ice formation. With no guarantee of avoiding freezing, the animals have then found a way to survive it. Their antifreeze glycerol serves a dual function. It lowers the freezing point, thus reducing the probability of freezing, but when or if freezing does occur, then the glycerol acts to reduce the damage caused by ice crystals. Indeed, glycerol is found in both freeze-intolerant and freeze-tolerant species, having a different function in each.

Freezing-avoidance by supercooling is of course also potentially adaptive. But only at *consistently* modest low temperatures. At even occasionally very low environmental temperatures, when freezing is highly possible, if not inevitable, then supercooling is dangerous because it could cause instant freezing and sure death. To understand why, we need to keep two things in mind. First, supercooled larvae, if seeded with an ice crystal, would freeze nearly instantly (the exact speed, in seconds, would depend on the amount of supercooling). Second, although some insects can indeed survive being frozen, they can do so *only* if that freezing proceeds *slowly*. Any insect—or any animal—that freezes instantly, as when supercooled, then also dies instantly.

Rate of freezing is important for cell survival because of a compartmentalization of ice crystals that relates ultimately to dehydration. During slow freezing the fluid *surrounding* the cells freezes first, because it is more dilute than the fluid within the cells. As ice crystals form extracellularly, they use up water and leave pockets of fluid of higher concentration. These pockets then act to osmotically *withdraw water from the inside of the cell*. In effect, *gradual* freezing results in the extracellular ice formations with a concomitant dehydration of the cell contents, so that no ice crystals are formed *within the cell*—ice crystals that would otherwise tear the cell organelles. Or, if ice crystals do form inside the cell, they are small and less damaging. Alternatively, the water may solidify into glass form—a type of liquid that is hard and in which the remaining water molecules are unavailable to form ice crystals perhaps because they closely adhere to the molecules of the cell structure. In contrast, during *rapid* freezing there is not sufficient time for the osmotic exchange of water (between intra- and extracellular) compartments, and so without prior dehydration the cell contents freeze, resulting in large, jagged ice crystals that shear and tear the cell organelles and membranes.

Let us return now to why northern *Eurosta* populations avoid supercooling by having a protein in the blood that *promotes* ice formation. They must thus have freezing-tolerance. Southern populations lack the protein that prevents supercooling and they can and most likely do supercool. Not having been selected for *freezing*-tolerance, they've perfected instead the mechanisms to *prevent* freezing rather than those that would help them survive it. The combination of antifreeze and supercooling is reliably *sufficient* to preclude ice formation in their environment. *Rate* of ice formation is no longer an issue when ice formation is unlikely, so no precautions for ensuring slow freezing (to promote cell dehydration freezing survival) are necessary.

A convincing demonstration that dehydration secondarily confers freezing tolerance can be found in the larvae of the African desert fly (*Polypedilum vanderplanki*), which periodically dry up in the temporary desert pools within which they live. The larvae are adapted to survive losing 92 percent of their body water, and such desiccated larvae are essentially immortal and can survive immersion in liquid helium (to -269°C), within 4°C or potentially at absolute zero, or 0°K, the lowest temperature in the universe. When rehydrated by dropping them into water they become "instant insects." They then again have narrow temperature tolerances, surviving only from 10° to 42°C. Are they alive before they are wetted? I think not. What they are is *potentially* alive, and I base that supposition on some oblique illumination from recent research in geology and microbiology.

About 250 million years ago—that's about 190 million years before the dinosaurs went extinct—there was an inland sea in North America. The ocean eventually evaporated where there is now the

Polyphemus moth pupae from
the field in winter have three-way
protection:
· tough shells (cocoons) that are
  impenetrable by most birds
· a camouflage wrapping of
  dead leaves
· biochemical protection to
  prevent death by freezing

New Mexico desert. It left salt deposits a half mile below the now sunbaked ground. Tiny pockets of ancient seawater became trapped in these salt deposits and from them microbiologists have reported isolating, and then growing (actively metabolizing and reproducing), novel *Bacillus*-like bacteria. Additionally, they found other microbes whose DNA does not match DNA of known organisms, so that these novel forms are thought to be the first, or among the first, ancient microorganisms that inhabited the planet. Bacteria had previously been isolated from guts of dead insects entrapped in amber for 125 million years, and the microbiologists reported that these bacteria could also be grown in culture, making it possible that living 125-million-year-old organisms had been found.

So fantastic are these results that they should be and are viewed with skepticism in the scientific community. However, I think they are fantastic, not because of magical processes new to science that we don't yet understand, but because of the enormous time spans concerned. (I'm personally skeptical, but will give my verdict shortly.) That is, if the bacteria had lain dormant for one, ten, or a hundred years, nobody would have blinked an eye. Even complex, multicellular animals can revive after being dried for a century: Six-legged primitive insectlike arthropods called tardigrades have been known to walk off after they had been inadvertently preserved in dried moss specimens in museums. What is remarkable is not that these organisms survive any specific number of years while in an apparently lifeless state. What is remarkable, to me, is that they can survive in that state for even one minute.

If even multicellular animals, such as insects with organs, muscles, glands, and nervous, reproductive, digestive, and excretory systems, can survive having all of those systems, each with its thousands of cells, being stopped and potentially disrupted, then how much easier it must be to stop and restart the nevertheless highly orchestrated chain of complex biochemical reactions in a single cell, or an even much simpler system, such as a virus? Viruses are not primitive organisms. They are inordinately elegant life-forms that are functionally reduced to a bare minimum required for growth and reproduction in their specific environment, the interior of cells. Their liveness is a complex series of chemical reactions that is within our grasp of understanding. It is at least theoretically possible for us to synthesize them from chemicals "off the shelf." A bacterium is admittedly a much more complex combination of molecules than a virus, but the same principles apply. If a fly or a tardigrade can survive being dried for a century (and presumably much longer), then a bacterium could be potentially immortal.

A DNA molecule could remain forever; if it is preserved in a saline solution and is not degraded, then time is irrelevant to it. Only the ambient conditions are vital. Given constant conditions, there is no scientific surprise if it survives or has a "shelf life" of 10 years, 10 million years, or 100 million years. I'm *emotionally* shocked and surprised that bacteria may have been preserved for 250 million years, but intellectually I'm less surprised that, if indeed preserved, they could still metabolize, grow, and divide when recovered. However, I do not subscribe to the idea that they were *alive* all this time. They were dead as a rock. And so is an insect that freezes in winter. It is not matter that defines life. Process, such as energy flow, does.

Research on insects has opened the amazing possibility, only broached in folklore, science fiction, and most recently even business, of freezing ourselves solid and later reviving. After being frozen the body is "dead" by most definitions (no movement, heartbeat, circulation, respiration, neurological activity) and there should then be no limit to the length of time a body can be preserved. Thankfully, no humans have been immortalized in ice. But freezing of human embryos has been in practice since 1984; some human embryos have been in storage for eight years before being implanted. There is no obvious physiological reason why they could not be stored for more than a century (see *Time* magazine, March 2, 1998), but I could think of lots of moral ones.

There is a whimsical story of the townsfolk in a village in the Northeast Kingdom of Vermont—an isolated backwoods area known for its cold winters—where the residents of one little village were said to avoid the awful winters by downing a few stiff drinks in the fall and then freezing themselves solid and then unthawing to resume an active life at an appropriate time in the spring. The movie *Iceman* similarly featured a man who had been frozen in ice some ten thousand years ago, who was subsequently thawed and brought back to life. Some people are willing to believe that these stories are in the realm of possibility, which is why a company (TransTime Inc. of San Leandro, California) provides "commercial cryonics and cryonic suspension services"; they will freeze anyone in liquid nitrogen and keep them frozen, presumably for the next millennium or beyond, for a "minimum of $150,000," at 2002 prices. Walt Disney is said to have bought such cold treatment rather than opting for cremation, despite the no-guarantee (of survival) clause. Currently, some "patients" have been "maintained" for twenty-three years. As far as I know, however, nobody has yet submitted to the treatment while they were still in the prime of health, which is when all the animals do it.

The potential implications of the knowledge of freezing-tolerance do not seem to be lost on the agencies that fund research. One researcher with whom I talked had at one time worked on frost-tolerance in insects. He told me that after switching to work on vertebrate animals, "they practically threw the money at me." I would probably have taken such money as well, if offered. But I confess to unease. The promise to make us immortal and the specter of frozen bodies in vats horrifies me. The unintended implications alone are all too obvious to need reiteration here. It is pure research, that which has no practical implication whatsoever, that enlivens the human spirit the most.

Even before I finished building my cabin in
Maine, I could see that it had potential. Bubo,
my tame great horned owl, chose to perch on the
rafters rather than out in the woods, where he or
she was harassed by the blue jays. Similarly in June
the hordes of bloodsucking blackflies and horse-
flies left off their hot pursuit as soon as I crossed
the doorstep. The cabin was a sanctuary for me in
the summer. When winter came to the Maine
woods, however, it suddenly became appealing to
the wild local fauna, and many adopted my haven
as their own.

Masked shrews and red-backed voles, my occa-
sional winter visitors from the subnivian zone,
were only transient visitors. In contrast, white-
footed mice took up permanent winter residence.
For some reason they find the cabin congenial.
But before I tell you more about them, I need to

describe and identify them. According to Mason A. Walton, the so-called Hermit of Gloucester, who in 1903 wrote about them in his book, *A Hermit's Wild Friends or Eighteen Years in the Woods:* "The white-footed mouse, unlike the house mouse, is a handsome fellow. He sports a chestnut coat, a white vest, reddish brown trousers, and white stockings. His eyes and ears are uncommonly large, causing his head to resemble a deer's in miniature. This resemblance has bestowed upon him the name of 'deer-mouse.' " (p. 118)

Deer mice juveniles have lead-gray pelage and white bellies. Unlike meadow voles or field mice, they also have long legs that allow them to bound like deer. There are, however, two species of closely related deer mice, and only one of them is the official deer mouse, *Peromyscus maniculatus.* The other is the white-footed mouse, *Peromyscus leucopus.*

Differences between the two species are subtle. One field guide I consulted indicated that in white-footed mice the tail is longer than the body, while in the deer mice it is shorter. But those that I measured at my cabin had tail lengths about equal to their body length. Only experts can distinguish the two, and the defining characteristic used for differentiating them is a molecular variation in their salivary amylase, an enzyme in their saliva that helps digest starch. Bill Kilpatrick, the mammologist I consulted, told me that mine were indeed deer mice, *Peromyscus maniculatus.* This information makes a difference to me, because here in the East, only *Peromyscus leucopus* is known to carry the Hanta virus, which is lethal to humans. (However, I'm not convinced that a virus capable of jumping from *P. leucopus* to *H. sapiens* would be incapable of transmission to *P. maniculatus.*)

Even with Hanta virus out of the picture, deer mice can be objectionable in a cabin, and in the winter they enter in droves. I can't blame them, though. The fault is mine. I should have used dry, nonshrinkable ceiling boards to foil these partly arboreal mice. Nor should I have used Styrofoam panels for ceiling insulation. I had not been warned that *Peromyscus* systematically shreds Styrofoam into chips. The chips drift down like unmeltable snow through cracks between the boards and fly up into the air when one tries to sweep them up. The mice, once inside, also raid one's dry goods, and use one's shoes, and bed, to hide them in. The Hermit of Gloucester, who lived before the age of Styrofoam, had dozens around him simultaneously. He was entertained by them, yet even he acknowledged, "A few mice for company on winter evenings would not be objectionable, but I draw the line when I am forced to eat and sleep with them." Relocating them, Walton learned, has little effect. One night he caught twenty-eight deer mice in his cabin and released them a mile distant. The whole crowd returned by the second night, noisily announcing their presence with the drumming of their tiny feet (a sound by now familiar to me). Deer mice, which utter no vocal sounds perceptive to our ears, use these drumrolls to communicate messages to one another—messages that remain undeciphered by man.

Deer mice are cute, and grudgingly I admire their liveliness and resourcefulness. They live full-time in the woods, where even without the handy materials of Styrofoam or crumpled sweaters, they manage to build excellent nests. At our home in Vermont, they have neither, and most stay outside. Recently at least one set up housekeeping inside the brain cavity of a moose skull that has long hung on the chicken house. (The skull is an eighteen-year-old memento. It came from the poached moose whose carcass attracted a crowd of ravens that started my study of ravens in the wild.) The entire nest in the cramped quarters of the moose skull was a solid ball of Rhode Island red chicken feathers no doubt gathered from inside the chicken house. On December 6, 2001, the skull was unoccupied, and I checked the birdbox in the woods adjacent to the raven aviary to see if the mice had changed residence. It seemed that the roomy log birdbox might be preferable to a moose skull, as woodpecker holes make natural nesting places for the mice. The birdbox consists of a small section of hollow log drilled through with a hole with one board nailed over the base and another board set on the top and secured with wire so that it could be opened.

I worked clumsily to remove the top. When I was finally able to look inside I saw the typical domed-over nest of a deer mouse. This one was made almost entirely of fur. Nothing stirred in the nest, so I started pulling out felted wads, when two *Peromyscus* immediately shot straight up and almost into my face and then bolted for the woods in long bounds. A third mouse poked its head through what was left of its nest, looking at me intently with its large black eyes. I replaced the cover at once and stepped back. The remaining mouse then poked its head out of the hole before running off as well.

No food was stored in either the skull or the birdbox, nor in any of the dozens of others that I have examined. Yet deer mice do cache their food. I have found their hoards of seeds not only in shoes in the cabin, but also under loose bark in the woods and in abandoned bird nests that were specifically domed over to hide the seeds (see Chapter 5). Why don't they store their food more conveniently, right in their own nests? I suspect it has to do with private property. Deer mice in the winter huddle not only with relatives, but also with nonkin. Not even deer mice are interested in working hard for an investment that others (especially nonrelatives) might reap as their own.

Deer mice have been intensively studied for more than a half century: everyone wants to know how they survive northern winters. Researchers are in agreement that deer mice don't hibernate. The relevant question then becomes how such animals weighing only about 20 grams each manage to survive *without* hibernating? As is usually the case when an animal is up against a difficult problem, it uses every trick available to solve that problem.

As we've seen with birds, amphibians, even insects, the key to winter survival is energy economy. To this end, deer mice employ several strategies, each of which has been well studied and documented.

Deer mice are nocturnal, and when not active, they retire to the snug nests they generally build in woodpecker holes and other tree holes. Captive *Peromyscus leucopus* kept at 25°C build deep nests in a day or two. At 30°C, however, they don't bother (Glaser and Lustick 1975). It costs energy to build a nest, but it yields energy savings in the long run by reducing the fuel costs of keeping warm. Even more energy is saved by huddling with others (Sealander 1952). Some individuals take it even further, by becoming torpid and reducing their body temperatures to near 20°C in the daytime. The various behaviors are adaptations, since deer mice from northern areas are more prone to enter daily torpor, build larger nests, and store food than those from more southerly areas (Pierce and Vogt 1993). Combinations of these several adaptive energy-saving strategies make a big difference in the winter, when energy supplies are often limited. For example, at 13°C, those mice that are nontorpid, nonhuddling, and nestless expend 2.5 times more energy per day than those that employ all three energy-saving strategies (Vogt and Lynch 1982).

Normally the torpor of the mice begins at daybreak and is over by late afternoon. Thus far, no one knows what cues the mice to enter or arouse from torpor. Curiously, they shift from carbohydrate to fat metabolism several hours before entering torpor (Nestler 1990), in a metabolic shift reminiscent of that which occurs in animals engaged in prolonged exercise.

Although the strategies of the deer mice act to save energy and thereby preserve their energy stores, winter adaptation in deer mice paradoxically involves being *able* to preserve energy expenditure. Normally the mice are active at night or at least during part of the night. At that time low temperatures cannot be avoided, so the mice must acclimate to them (Sealander 1951). Key to that adaptation is the adding of new red blood cells that have a higher hemoglobin content (Sealander 1962) to the circulation. As a consequence, the mice can increase their metabolic rates and hence their ability to tolerate and keep warm by shivering at low air temperatures. By living in my cabin and in other human dwellings during the winter, deer mice exhibit another manifestation of their energy-saving strategy: Temperatures in the cabin are not as low as those outside the cabin, fuel is plentiful, nesting material and sites are conveniently available, and so the mice can afford to remain nontorpid and active longer.

Deer and white-footed mice contrast with another species pair of similar mice common in New England, who like *Peromyscus* are not permanent residents of the subnivian zone. But these two species, the meadow jumping mouse (*Zapus hudsonicus*) and the woodland jumping mouse (*Napaeozapus insignis*) never enter our houses. They are deep hibernators and stay outside. Neither species has cheek pouches for carrying food, like some other hibernators that store food (pocket mice, kangaroo rats, chipmunks, hamsters), and they do not store food. Instead, they fatten up prior to hibernation. Like *Peromyscus*, both species have yellow-gold pelage and a blackish stripe on the back. When in a hurry, say, making an escape, they move in successive leaps of about four feet long each, aided by powerful hind legs, their long, white-tipped tails extended out behind. They are seldom seen, although Carolyn Sheldon who studied both species near Woodstock, Vermont, from 1934 to 1937, reports that the meadow jumping mouse was familiar to the local farmers.

Sheldon's (1938a,b) study involved capturing and marking numerous individuals of both species near Woodstock, Vermont, to determine their home ranges. She also tried to raise them in captivity. *Zapus* never mated in captivity, and a pregnant female that was captured and gave birth to seven young paid no attention to them despite their cries. By three days, they had died. *Napaeozapus,* on the other hand, mated readily in captivity and produced many litters, although Sheldon reported that the mother "nearly always" destroyed (ate) them within twenty-four hours. In captivity both species fed on seeds, berries, powdered milk, insects, and their own young when available. Obviously their environment in captivity was not like that to which their psychology is evolved. Either a key element was missing or a harmful one was present.

In the field, *Zapus* builds aboveground summer nests of closely woven fine grass and dry leaves, with a small difficult-to-find entrance to the side. In the fall the mice leave these summer nests and burrow into the ground, where they hibernate. In captivity, mice began digging and making underground nests for hibernation during the last part of September, when temperatures dropped to 5°C. At first the mice's periods of torpor lasted only a few hours: They awoke frequently, for a few hours and then a few days, before settling in to prolonged torpor of more than two or three days at a time. In Ithaca, New York, hibernating mice have been dug up as late as the end of April (Hamilton 1935), and Sheldon (1938a) even found them in torpor in her live traps in the last part of May. *Napaeozapus* also "became sleepy" in September and October, and like *Zapus* they then also left their aboveground nests and built new underground nests for hibernation. Like deer mice, jumping mice find new quarters in the winter. Theirs, thankfully, are not with me.

The mice apparently get inside through tiny holes and by chewing and pulling out the oakum fiber plugs between logs. In this they may be aided by the stronger red and flying squirrels, which also pull out the oakum for use in their nests. Whether independently or working together the mammals ultimately provide the main means of entrance for a crowd of insects.

The crowd is always snug inside our Maine cabin, come winter. It consists mostly of cluster flies. According to Harold Oldroyd, who wrote the fly bible, the 1964 *Natural History of Flies*, there are several species of these robust members of the genus *Pollenia*. Most of them are several times the size of the more familiar housefly. *Pollenia* are calliphorid, or "flesh" flies. The larvae of North American native species eat the flesh of dead animals, but the most common flies in the cabin, *Pollenia rudis*, were introduced from Europe and their larvae parasitize earthworms; they eat them alive from the inside out. Big and bristly, these flies are not handsome, like the shiny metallic green-and-blue native *Pollenia,* which never enter the cabin. Already in the fall the *Pollenia rudis* perch in crowds on the logs outside the cabin and sun themselves. When it starts to cool, they slip through the cracks. By November most have made their way inside, but at that time they remain unobtrusive unless I build a roaring fire in the woodstove. Then within minutes they come poking out of the cracks and crevices, and if it is still daylight hundreds or thousands gather in buzzing masses at each of the eight windows making a collective hiss. They apparently perceive the warmth as the return of spring and take it as their signal to try to leave by flying directly to the light at the windows. If given a chance to exit, at least they do not try to stay on, unlike some other previously mentioned guests. Even on the coldest days they rush out instantly when I open the windows wide but they can fly only a short distance before the cold grips them and they plummet immobile to the snow. I've captured hundreds of the flies at my windows, painted them with dabs of red paint, and then released them outside to see if they return. When I used my Dustbuster a few days later to vacuum them up by the cupfuls at the windows, I got mostly new ones but there were some returnees.

I've become curious about their overwintering adaptations because I haven't found any of them in the wild. Temperatures can dip to -30°C outside (and the cabin is unheated for most of the winter). Some of these flies that I've brought into my lab in Vermont and subjected to -20°C were frozen solid and dead in minutes. But at -10°C, still a bitterly cold temperature, others did not freeze and survived; within seconds of being warmed they crawled about and again flew as vigorously as minutes before. I suspect they might survive by supercooling, and the very dry environment in the cabin is ideal for supercooling.

Overwintering in the super-cooled state can be a dangerous gamble, because contact with ice crystals can provide nucleation sites (places for ice to start forming) and the whole animal can then freeze solid in seconds, meaning certain death. Many insects survive the entire winter while supercooled, but in order to do so they require

Alaskan wasp queen hibernating by hanging to avoid ice crystals and maintain supercooling.

overwintering sites where they can avoid any contact with ice. For example, the temperature at which Alaskan green stinkbugs crystallize into ice (and die) is near -2°C when they come into contact with snow, but they remain unfrozen and alive down to near -18°C when kept dry (Barnes et al. 1996). The overwintering queens of yellowjacket wasp (*Vespula vulgaris*) in interior Alaska also supercool. The wasps isolate themselves from contact with ice by attaching their mandibles to the undersides of leaves or leaf litter and then hang suspended through the winter. The supercooling points of these free-hanging wasps as measured in the laboratory decreased from near -10°C at the beginning to near -16°C in late winter, whereas temperatures in their hibernacula were always higher. Thus, these freeze-intolerant insects suffer little winter mortality from freezing. Curiously, the queens of another wasp variety, the white-faced hornet (*Vespula maculata*), from South Bend, Indiana, are freeze-*tolerant*, and for overwintering they produce ice-nucleating factors in the blood that *promote* freezing, to prevent supercooling (Duman and Patterson 1978). Unlike those of the yellowjackets, their nests are not underground. In Maine and Vermont, where I live and work, their big gray paper nests are suspended in trees, but these are empty in winter. I've not yet found any of their overwintering queens.

There are those insects that can't avoid moisture and therefore can't supercool. They include the grubs of long-horned beetles (Cerambycidae) that feed in live, moist wood. These "sawyers," so-called because they leave telltale piles of sawdust wherever they burrow out of a log, chew through wood in the summer with their hard sclerotized mandibles, making a sound like someone using a cross-cut saw. (The adults, curiously, make a squeaky cry when you hold them, using a little scraper in the head-thorax junction.) Larvae that I dug out of the wood in summer and placed in -10°C conditions quickly froze into solid blocks, and they were then dead. The dead white grubs, when thawed, turned brown in a day. By winter the adults have long died, and the overwintering larvae are immobile and silent in their wood galleries. They now survive the low ambient air temperatures of winter, well below -20°C, and they now do *not* freeze into solid blocks because they then have antifreeze in their blood.

The logs of our cabin do not contain any sawyer grubs because when I built it I peeled the fresh logs so that they would dry quickly and become unsuitable food for them. However, the logs did eventually contain a large colony of carpenter ants. Bad news, because they are permanent residents, once ensconced. They produced huge piles of sawdust in far greater amounts than beetle grubs would have produced. I feared they would hollow out the logs and cause them to collapse. After a few years of their presence there, with no end in sight, I was, in desperation, about to hire an exterminator when, in the summer of 2001, a huge phalanx of red ants (*Formica subintegra*—which are normally slave-raiders of the Formica ants) came in, waged war, and within one week heaps of still-wiggling dismembered carpenter ants (*Camponotus*) and carpenter ant parts were strewn inside and outside the cabin. The raiders totally eradicated the carpenter ant colony, and then went back to their huge nest in the nearby field.

The carpenter ants are able to use the dry wood, because unlike the stationary beetle larvae, they are highly mobile and bring water into their nests to moisten the wood, if need be. I've found them living in still-upright balsam fir trees in winter and, since ice crystals are mixed in with the masses of comatose ants, they need some other strategy besides supercooling to survive winter.

Both the ants and the beetle larvae spending the winter inside tree trunks endure temperatures close to those of the ambient air. I've brought both into the house and warmed them up, but unlike flies, which spring to life almost immediately when warmed, the ants and beetle larvae *seem* stone dead, even after being warmed. Only after a few days at room temperature do they gradually show movement, eventually resuming full activity. But when I have taken these revived ants and beetle larvae and stuck them out in the cold from whence they came, they died quickly. Clearly their survival is dependent on their antifreeze-induced torpor. The ants and beetle grubs, when cold-adapted, contain large amounts of glycerol or other sweet-tasting antifreeze (I have not tasted the flies), which prevents ice crystals from forming in their bodies and probably pickles them into inactivity. It takes a long time at elevated temperatures to get that antifreeze out of the blood.

The other winter cabin guests—I'm thinking of three species in particular—are beautiful and more benign in habit. When the cabin is heated, for example, they don't have the annoying *Pollenia*'s tendency to hover around the bed light and then, when it's turned off, to dash under the covers and buzz there rudely.

The first of these three species, the mourning cloak butterflies (*Nymphalis antiopa*), usually remain in crevices outside and only rarely make it into the cabin. In the fall I commonly see one or two fluttering under the cabin roof. The second species, the multicolored Asian ladybugs (or ladybird beetles), has arrived here in numbers only in the past few years. These occupy the cabin by the thousands in some years, and by only dozens in others. First imported from eastern Asia by the U.S. Department of Agriculture in the late 1980s as a biological control on pecan aphids, this species (*Harmonia axyridis*) was later introduced to other areas, including the Northeast, to control similar pests. It has since spread far and wide on its own, to most areas of the United States and to parts of Canada. Like many species of ladybirds (there may be some four hundred ladybug species in North America and three thousand or so worldwide), these beetles all have handsome black coloration. Those ladybugs endemic to North America species have specific color patterns, but in the Asian species, virtually every individual is uniquely colored. Each bug's background hue can be deep red, or orange, or yellow, and it may have no spots, tiny black dots, or spots that coalesce into black bands. (Another species that very rarely overwinter in the cabin is uniformly coal black with one red spot on each wing cover.) The colors serve as warning to stay clear; multicolored Asian ladybugs secrete a foul-smelling fluid when crushed. But even certified bug-haters find it hard to want to kill ladybird beetles. Besides being prettified in handsome colors, they have soft rounded curves, little legs, and petite feet. They are about as cute as any bug on the planet. Adorable, even. But cuddly, they are not. We found that out after they moved into our Vermont home in the fall by the tens of thousands (literally), and then stayed for over six months—despite all our best efforts to try to evict them.

A sample of multicolored Asian ladybird beetles from my window.

Our family made a most intimate acquaintance with these beasts during the winter of 2001–2002. A few pesky forerunners had appeared in our Vermont home in previous years, when they were an almost welcome diversion from the cluster flies. But that winter, the cluster flies paradoxically were almost absent, while the beetles staged an invasion. I don't know how, or why. I tested one batch in the freezer compartment (at -14°C) of the refrigerator, where they readily (well, 61 of 148 subjects) survived.

Spiders are fortunately not alive in my winter cabin. Here is "Charlotte" dead and shriveled on a beam, next to her eggs, which do survive.

I had over the winter opened the bedroom window on numerous occasions to then brush them off the panes and the walls and throw them out by the hundreds. Still there were always plenty left at night congregating on the reading lamp. After we shut it off, the bugs then settled into our beds. Being dehydrated by the dry indoors, they not infrequently tried to crawl into our eyes at night, looking to suck up some of our moisture or to nip at our skin. By Easter, they started to come out of hibernation in force. We then had an infestation by the thousands, and as a last resort my wife Rachel tried to vacuum them up with a Dustbuster. Like riled-up skunks, they released their toxic chemical defensive secretions. Rachel is a biologist and the paragon of tolerance toward creatures. She is enthusiastic when the kids bring in spiders, earthworms, slugs, millipedes, sow bugs, and centipedes. But by April 2001, after another go at them with the Dustbuster, she declared *Harmonia axyridis* "disgusting." I didn't argue.

My grudging tolerance for the beetles stems mostly from their predatory habits. In addition to pecan aphids, the multicolored Asian ladybugs and their larvae feed on plant-sucking insects such as the woolly adelgid, which is decimating hemlock stands from Virginia to New England. Over its lifetime, a single multicolored Asian ladybug can devour an estimated 600 to 1,200 aphids. In the 1960s the adelgid, too, was introduced from Asia (although it arrived inadvertently, probably on a nursery plant). Twenty years later it had become a serious problem. A hemlock tree attacked by adelgids invariably dies. So far, I have not had a problem with adelgids on my hemlocks.

The third insect species that is a regular if not abundant visitor is the green lacewing (family Chrysopidae). The light, bright green of the lacewing extends to its four large wings, delicate membranes stretched between a network of veins. Lacewings have a certain aura. But this is lost on aphids. Lacewing adults as well as their larvae, commonly called aphid lions, are ferocious predators like their relatives the antlions, whose larvae build deadly sand traps that capture ants. In contrast to all of the other house and cabin winter crowd, I seldom find more than a half dozen, but I commonly see them (unlike any of the other cabin occupants) under loose, dry bark of trees in the winter woods. They are rare enough guests in the cabin to be a treat.

Some deliberately bring houseguests into their dwellings for the winter. Our son Eliot and I found some tiny pale yellow ants overwintering in nest chambers under stones. In these chambers, attached to roots and rocks, we saw chalk-white blobs that, on close inspection, revealed themselves as aphids. Of course these aphids, close relatives of the dreaded adelgids, are there because they make themselves useful (to ants). They secrete sweet honeydew in the summer when the ants put them back out to pasture. And when they have finished milking them of honeydew in the fall, they tuck them safely back underground for winter storage into their chambers where we found them. They perch there all winter without, most likely, ever moving and being a nuisance.

The diversity and abundance of winter wildlife in my cabin is not necessarily enviable. Anyone can be similarly blessed. By allowing a few openings, I now play host to the good, the bad, the beautiful, and sometimes to the useful.

One of my *fondest childhood memories is of the* bats on summer evenings. As reliable as the swallows in the barn and the bobolinks in the hay-fields, I saw bats zigzagging around the barn in the evening. They fluttered over the fields and close over the water surface of nearby Pease Pond as we fished at dusk for white perch. I don't see them much now and I miss them.

My father was a bat enthusiast (in addition to his passion for ichneumon wasps and his interest in birds not to mention his pet Bulgarian weasel), and on some evenings we went out to hunt them with his shotgun. My mother, his preparator on many expeditions, skinned and stuffed them. Each species had its own flight signature in the way it fluttered or zigzagged, and the habitat where it could be found. It was challenging to learn about bats, and exciting to see them. His collection of perhaps two dozen species, which is displayed in glass-covered cases, ended up at the Bates Museum at the Hinckley–Good Will School in Maine. When I last saw those bats there in 1999, I was saddened. Not because we had killed them—their deaths, of course, brought us awareness and possibly knowledge. Rather, I was sad for the deep ignorance of those who have never seen, handled, or learned to appreciate bats, which is in part responsible for their population crashes all over the continent. Almost nobody sets out to do deliberate harm. Most evils happen inadvertently, through not knowing and uninformed notions.

At that time I did not wonder much about how bats spent the winter, nor did I (or anyone, until twenty-three years later) have a clue or care about the winter whereabouts of the monarch butterflies (*Danaus plexippus*), whose familiar yellow-white-and black-striped caterpillars fed on the milkweed patch next to the barn. I picked up caterpillars, fed them milkweed and raised them to adulthood in my room. I did not suspect that butterflies and bats had anything in common, nor that the continued survival of both depends on precise temperature regimes of their winter world, at pinpoints of the globe hundreds and even thousands of miles removed from the isolated Eden of the Maine woods, where they seemed part of the landscape. The official response of "protecting" these animals by making it illegal for curious kids to handle or collect them assumes that *everyone* wants to do it. By that logic one could just as well make it illegal to *not* handle wildlife, because *some* get enlightened by contact with it. Personally, I think that is ultimately more useful than everyone being distanced from it. Contact should be encouraged.

The monarch butterfly summers throughout the United States and into southern Canada, which is the northern limit of its food plant, milkweed. There are two major populations of the monarch that are separated by the Rocky Mountains. Those west of the Rockies migrate to the California coast in winter. There they overwinter in about forty colonies, including well-known ones at Muir Beach, Santa Cruz, and Pacific Grove.

For a long time it was not known where the eastern population spent the winter. In 1937, zoologist Fred A. Urquhart and his wife, Norah, suspecting that the butterflies migrated, started gluing tiny tags onto the wings of thousands of monarchs at their home base in Toronto, with instructions to send recoveries to them. By mapping recapture sites over a number of years, they were able to reconstruct the butterflies' flight lines and determine that they were migrating all the way to Mexico to overwinter.

We now know that the east-ern population extends from the eastern slope of the Rockies all the way to the Atlantic seaboard. Most of this popula-tion migrates south in the fall, with individuals in it traveling up to 4,500 kilometers to over-winter in twelve extraordinarily small patches of pines and firs

Monarch butterfly.

in the Transvolcanic Mountains in the Mexican state of Michoacan. The butterflies overwinter in these mountains at an altitude of 2,900 to 3,300 meters (9,500 to 11,000 feet) at preferred sites that have cool yet not too cold temperatures, high relative humidity, and little wind (Brower and Malcolm 1991). At one large colony where more than 14 million monarchs congregate in about 1.5 hectares—about 4 acres or less than a hundredth of a square mile—temperatures ranged from 5.6° to 15°C, near the butterflies' threshold for shivering to get ready to fly. It is here, at these sites, that the vulnerable heart of the North American monarch population resides and spends most of the winter in torpor.

To keep the spectacular migration spectacle alive requires protecting the cool forests that promote the torpor and extend the insects, energy supplies. It is a sobering thought that most of the population of eastern North America could be wiped out by an irresponsible woodcutter with a chain saw. Continual survival is only as secure as the weakest link, of the innumerable links—thousands—to the existence of any species. In the long term, however, it is not only the roosting groves that are critical. Given environmental change, such as the global warming that is now melting the glaciers all over the world at a phenomenal rate, the alternative and as yet unused potential roosting sites will become important in the future.

The larger the aggregation, the less the individual butterflies in it risk predation. However, a main reason the butterflies use specific overwintering sites is that they can maintain the low body temperatures there required to maintain energy balance while at rest for three months of virtually no feeding (Masters, Malcolm, and Brower 1988). On average the butterflies' body fat reserves are such that upon entering their hibernation site, they should last about ninety days at $15°C$ (ibid.). On the other hand, if the body temperature of resting butterflies were $30°C$, then their rate of resting metabolism would be high enough to exhaust their fat reserves in fewer than ten days. In addition, they would likely dehydrate.

It is also important to note that the migratory behavior—restlessness, flight direction, and maybe duration and destination—has *evolved*. The butterflies won't just stay at the first cool spot they hit, because it could become hot, or too cold. They rely on the *genetic*, or long-term, experience of the race. Hence the importance of *specific* overwintering sites that have proven to be safe for their ancestors.

By February as the days get longer and the monarchs' critical photoperiod of 11.3 hours is passed, the hibernating butterflies can again become reproductively active. There is, of course, nothing magical about 11.3 hours of daylight per day *as such* for reproduction, except for monarchs. Like overwintering sites, that specific time represents an evolved memory of the race. It is the photoperiod that in *their* evolutionary history has proven to be the best, for them to be prepared to be active. After this photoperiod is experienced, the butterflies simply wait for the next cue—temperature. Rising temperatures trigger a massive mating response in the colony (Brower et al. 1977). After mating (another cue), the females then migrate northward and eastward back into the United States and southern Canada. Along the way, they respond to the scent of milkweed, a cue for laying their green eggs on the newly emerging plants.

For a long time there was controversy on whether the monarchs colonized their entire eastern range with the first spring generation leaving the Mexican overwintering sites, or whether the northward march was achieved in steps, by successive generations. Thanks to the fact that milkweed contains cardenolides (to us nauseating chemicals that are also heart poisons), this issue has now been resolved: It requires up to four generations for the monarchs to reach their northernmost breeding grounds.

The method to generating the above answer exploited the fact that different milkweed populations contain specific chemically distinct arrays of cardenolides. The monarchs ingest these poisons as caterpillars, store them in the pupa and transfer them into the adult butterflies where they serve as a chemical defense against predators. By extracting these poisons from a butterfly, one can get a chemical "fingerprint" to match that found in the milkweed of different areas. For example, the overwintering monarchs in Mexico and those on the early leg (March and April) of the migration had cardenolides that are found in milkweeds that are not found in Mexico or in the southern United States. Thus by such biochemical sleuthing it could be deduced that the overwintering monarchs and the early spring migrants were derived from caterpillars that had grown up the previous summer in the northern United States and Canada. In contrast, the butterflies collected in May and June in North Dakota had cardenolide fingerprints matching those of milkweeds growing only in the southern United States.

Every fall I now eagerly and admiringly watch monarchs, our most conspicuous insect migrant. Day after day in October, the handsome orange-and-black-striped butterflies flap and sail lazily over the sunny fields, the woods, and water, all flying individually yet all heading in a southerly direction. In the evening they stop and gather on the purple New England asters to sip nectar, and in the morning they bask and shiver, rise into the air, and resume their journeys.

Individual butterflies tagged in Canada in the fall have been recovered thousands of miles south. By winter most of the eastern population has settled into their winter mountain retreat near Mexico City where their great-great-grandparents had been before them. It is a destination they seek out with incredible energy expense, never having been there, nor knowing where they are going. By the tons, in a spectacular shimmering orange display, they festoon the trees.

The million-dollar question is: Why do monarchs bother? Why don't they all hibernate and stay north, as do most butterflies? Like most questions that relate to history, this one does not have a simple answer: because history, especially evolutionary history, is never just one thing acting in isolation of everything else.

The monarch butterfly is a member of the family called the Daneidae, a tropical group. Monarch's relatives live in the tropical lowlands of New Guinea, and they are prominent in the American tropics as well. The ancestors of the present-day monarch butterfly, *Danaus plexippus*, were presumably also adapted to a tropical climate. Like most butterflies, they are predisposed to disperse when the larval food runs out. Not being able to survive freezing temperatures, there was then strong selective pressure for dispersion in a specific rather than a random direction; many dispersed in the wrong direction but all of those individuals left no offspring to pass on their trait. Only those who lucked out by flying southward survived; thus an evolutionary direction and destination was born and then grew.

At the same time they were evolving directional dispersal, monarchs had to surmount another problem. On their annual southern treks they had to cross into the hot, arid environments of the Mexican deserts. Here there was little chance for these strong fliers to refuel, and facing a long interlude until feeding was again possible, they must have experienced intense selective pressure to conserve energy. Partly by chance, some individuals probably ended up in the mountains, perhaps blown there by tropical updrafts. I once saw aggregations of insects on Mount Meru in Tanzania, and all of them were torpid. The cool mountain air had reduced the insects' rate of energy expenditure to such an ebb that they could not even fly. They were now in cold storage and along with that went a reprieve from needing to feed, until they were again warmed.

It seems probable that, given an insect's high reproductive rate, those few monarchs that took a flight path that somehow landed them in cold storage for the duration when no food was available would have a huge selective advantage over those that would have exhausted their energy supplies by staying in the heat in an environment with little food. By sifting for survivors, evolution selected where the population would overwinter. Generalities such as these should, given the right circumstances, also apply to other animals, including Lugong moths and bats.

In Australia, the Lugong moth (*Agrotis infusa*) also migrates to cool mountain areas where it clusters in large numbers (and where it was once an important food for Australian aborigines). The principle of the moths' migration is the same as that of the monarchs'—to conserve their fat reserves during a long quiescence—but rather than migrating to escape freezing, they migrate to escape *hot* conditions.

In bats, we are given a fine example involving escape from both low *and* high temperatures. Bats are, like the Daneidae and also the

Hominidae, animals of the tropics. Those that live in the north are outliers (as are *Danaus plexippus* among the Daneidae, and *Homo sapiens* among the Hominidae). Like us, bats are now able to live in the north, not because they tolerate freezing, but because they manage to avoid it. Like monarchs, many bats migrate, but their ability to do so leaves them much more leeway as to destination.

We tend to mostly think of migration as north-south movement, but migration can be in any direction whatsoever. Blackcap warblers from central Europe, for example, have traditionally migrated south, into Africa, in the winter. But within several decades a part of their population has evolved, by natural selection, to fly east-west instead, wintering in Great Britain where the weather is milder and bird feeders have become available. Similarly, northern bats of many species also migrate to where they can keep in energy balance. But that energy balance is achieved without feeding. Like the monarch butterflies, they migrate to cold storage environments where they can both conserve their fat reserves and not be endangered by freezing. With many bats, that means overwintering in caves.

If no feeding is possible for months, then just any cave won't do. Cave temperatures can't be below about 0°C, or else the bats risk death by freezing or energy exhaustion by shivering to prevent freezing. At the same time, cave temperatures can't be high if no food is available outside, because then even the animal's idling or resting metabolism would eventually exhaust their fat reserves. In general, each 10°C rise in body temperature doubles the rate of resting metabolism (i.e., resting energy expenditure).

In the south, some bat populations migrate north where cave temperatures (and lowest possible body temperatures) are low enough for them to remain in extended torpor (McNab 1974). Few bats are able to hibernate at cave temperatures above 14°C. The exceptions are very small bats and those that don't cluster, and thus enhance their ability to cool.

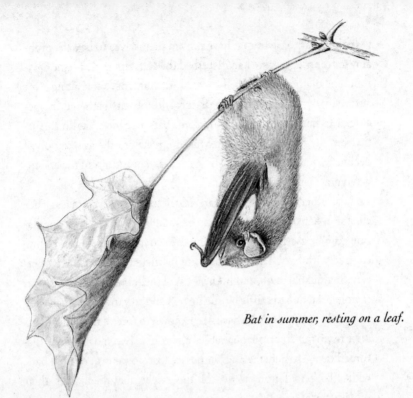

Bat in summer, resting on a leaf.

A bat entering a cave for the first time cannot know beforehand if the temperatures within will be suitable to maintain a positive energy until the end of the winter, any more than monarchs can actively choose specific mountain retreats where their energy balance will come out just right by spring. However, if the animal ends up surviving, then conditions *were* suitable—the energy balance came out favorable, and the bat's survivors will likely return to that site the next winter. This is what bats do. It's not only natural selection that's operating on bats to survive winter; there is also selection for caves that *become* bat caves; in those caves that are not suitable, the populations never build up. Conversely, once having built up, they decline if conditions become unfavorable.

Bats are long-lived animals that learn by experience, returning year after year to caves that have proven themselves to be safe, probably for centuries. Little had disturbed the constant and specific environment of their traditional caves until humans came along. Not surprisingly, therefore, human disturbance of bats gathered in specific caves has been a big factor in some bats' declines. Merlin D. Tuttle from the Milwaukee Public Museum reviewed the association of decline with disturbance by people in caves, specifically of the endangered gray bat (*Myotis grisescens*).

Like other bats, gray bats are restricted to specific caves. They cluster in densities of over 1,800 per square meter, and cave populations can be assessed by estimating the square meters of cave ceiling covered with bats. Their colonies are restricted to fewer than 5 percent of available caves, and in these caves the human disturbance has mainly been due to traffic by spelunkers and to vandalism, including by health authorities who have been known to torch a cave full of bats after receiving an erroneous rabies claim. The two most heavily disturbed caves in Alabama and Tennessee lost 90 percent of their bats, while the population in five of the rarely disturbed caves there remained stable.

In an attempt to stop the sometimes catastrophic declines, cave entrances were in many instances altered to restrict or limit human intrusion. Ironically, however, the results of these well-meaning measures were mixed; sometimes the population recovered, but in other cases improperly constructed gates resulted in the loss of entire colonies. Potential causes are illustrated in the endangered Indiana bats, *Myotis sodalis*. Female Indiana bats live nearly fifteen years and males slightly less (Humphrey and Cope 1977). Reproduction is slow. Females have their first pup at age two, and after that have only one per year.

This bats' summer range covers most of the eastern United States, but about 85 percent of the population winters in seven caves; and half of the population can be found in just two. Since gaining legal protection in 1973, winter populations of the Indiana bat have decreased by about 28 percent until 1980–1981, and additionally by 36 percent in the next decade. A recent study by four researchers (Richter et al. 1993) from four different museums suggests that the bat's perplexing declines were due to modification of cave entrances. For example, from in the early 1960s when entrance modifications were made, to the early 1990s the bat population of Hundred Dome Cave in Kentucky declined from 100,000 to 50 bats. When the entrance of the Wyandotte Cave in Indiana was constricted by a man-made stone wall, the bat population declined from 15,000 to 1,400 bats by 1957, in twenty-five years. One thousand to 2,000 bats continued to overwinter there until 1977, when the stone wall was removed. Immediately after that the population rebounded until fifteen years later, when it was back up to nearly what it had been originally. The researchers ultimately concluded that the modifications of the cave entrances had their main effect of restricting airflow so that temperatures inside had become higher. Those at Hundred Dome Cave, where there had been the most precipitous decline, had increased from 4° to 6° to 11°C. As a result, these hibernating bats, normally found in conditions of 4° to 8°C, were insufficiently cooled.

In this species the body temperature during hibernation is essentially identical to that of the air, from -3° to 30°C (Henshaw and Folk 1966). At the very lowest end of this temperature range the animals became aroused and exhibit mild shivering, to heat themselves up slightly above air temperature. (In a closely related species, *Myotis lucifugus*, the individuals cannot arouse from such low temperatures and they freeze to death at near -5°C.) However, the main danger to the Indiana bats in their traditional caves is not quick freezing, but slowly starving to death in temperatures above 10°C when their elevated resting metabolism eventually exhausts their fat reserves by the end of the winter.

To test the latter hypothesis, biologist Andreas Richter from Earlham College in Indiana, and three colleagues, compared body weight losses of bats in two caves of different temperatures. The body mass loss was 42 percent more rapid in bats roosting at higher temperatures, and the dead bats were emaciated bats. The mortality inferred by the conditions in the altered Wyandotte Cave, from 1953 to 1978, should actually have been high enough to eliminate the entire population at that cave. But bats attract each other, and the apparent stabilization of the cave population of one to three thousand individuals was more *bad* news than good. It was created by influx of other animals from elsewhere. The cave had become a sink, a death trap. The deleterious effects of just 5°C higher temperature extended far beyond that of the cave itself. The fate of bats and butterflies is balanced on the winter world to which they are adapted.

All kinds of creatures form tight-knit societies
in winter, even those that don't crash a cozy cabin
and even those that don't need to seek warmth.
One November day some years ago I squatted
down on the freshly fallen but already matted
leaves in the woods. Within seconds I detected
the unmistakable odor of stinkbug. Digging under
the leaves, I found dozens of them massed
together, presumably settling in for hibernation.
I'd disturbed them in their bivouac, and they were
giving off their foul-smelling defense secretions.
I did not need to taste them—I knew they tasted
as bad as monarchs (but not nearly as bad, according to this gourmet, as a mass of overwintering
spider eggs!).

It is not only the stinkbug that smells or tastes
foul. Almost any insect that is brightly colored
(except some mimics of them) is sure to do the

same. The ladybird (or ladybug) beetles that aggregate by the thousands both in my Maine cabin and my Vermont home in the fall, when hoping to stay the winter and when unduly disturbed, put out a foul burnt-rubber smell that is overpowering. Some species of these beetles of the family Coccinellidae aggregate by the millions, and in California and other areas of the western United States where they mass up under rocks or at the base of trees up in the mountains, they are scooped up in buckets and sold to gardeners for aphid control. The reason for aggregating in both stinkbugs and ladybird beetles is likely for the purpose of massing stink power. That is, if you want to be associated with stink, make your own, look like a stinker, and better still go where others stink like you.

Aggregating in winter brings animals, at least some snakes, other advantages. One of the most amazing snake aggregations are those of the Manitoba red-sided garter snakes (*Thamnophis sirtalis parietalis*). Like stinkbugs, garter snakes give off a foul smell when you tread on or otherwise molest them. Each fall writhing masses of these snakes pack themselves like live spaghetti into specific crevices—ten thousand in a single depression of a few cubic feet—in the rocks of a barren region near Winnipeg. The snakes spending the winter at these spots avoid freezing and gain protection. While they're all together, they perform another primal function just before dispersing in the spring.

Males emerge from the rocky crevices before females and then wait at the periphery to intercept the females. As soon as a female emerges from the den, she is enveloped in a ball of dozens of suitors. Curiously, some of these males mimic females, and are mistaken as such by other males (Shine and Mason 2001). Their behavior just doesn't make sense (yet), but with more information I trust that it eventually will.

Aggregation behavior has its share of such mysteries. Some years ago I received a letter from a man in Alaska who wrote of seeing a communal crow roost in winter woods, where "the ground was littered with fighting crows who were murderously hacking one another to death." He had never seen anything like it. Neither have I. Nor could I make any sense out it, no matter how hard I tried to twist the scenario into a logical possibility.

Communal bird roosts are not closed societies. In the Maine woods, my colleague John Marzluff (who worked with me on ravens for three years) routinely introduced long-captive ravens into established communal raven roosts, and these new birds showed no hesitation in joining and were immediately accepted by the group. The next morning they then followed the roost occupants to the crowd's feeding place, such as a deer or cow carcass. There was never any exclusion of strangers. Communally roosting birds even tolerate other species. Crow roosts in the Old World sometimes contain magpies, jackdaws, and ravens. Although on the whole, roosts of corvids in North America tend to be species-specific, large roosts of icterids (blackbirds) near Burlington, Vermont, sometimes contain red-winged blackbirds, common grackles, and cowbirds. There seems to be almost no limit to the numbers of birds tolerated. There are reports of crow roosts in the western United States containing several million individuals.

Nothing in the scientific literature about roost behavior would allow for group fighting, and birds that fight and are territorial during the breeding season relinquish their antagonisms in order to be in roosts. What could there be to fight over, since the occupants join up because they ultimately need each other? Also, when social birds do fight, they don't hack themselves to death. (So how do I explain that letter from the man in Alaska? Bear with me a bit and I will try.) Crows sometimes act aggressively, but from what I have seen it has always been a group of them attacking another individual; I have never seen more than one intended victim at a time. It seemed bizarre that crows would wage the equivalent of a war at a communal roost,

AGGREGATING FOR WINTER

231

and that they would stay and fight there, without escaping, until bloody mutual destruction had set in. In short, I did not believe what I read because it did not fit into my previous knowledge, preconceptions, or experience. It just seemed to be another one of those bizarre reports that I hear all the time that are almost invariably the result of false identification or faulty observations. I would therefore have banished the report of the murderous crows from my mind had it not been for some observations of crows I made one evening in the city of Burlington, Vermont.

I have watched crows every winter for the past twenty years as they come to the Burlington area each winter. At dusk they start to form their communal sleeping roosts, which number in the thousands. They come flying into their roosts in endless long strings or queues flying high, forming diffuse, gray cloudlike aggregations against the snowcap of distant Mount Mansfield. Just when I think the last of them have arrived, I see many more behind, in what seems like an endless stream. All of them converge on one darkening spot near or in the city.

As spectacular as these flights to the evening roost are, to me the most notable thing about them is where the birds settle for the night. The roosts are never in the forests where crows commonly nest in the summer, nor are they in the pines along the fields that crows like to inhabit. Curiously, the city center itself now seems to be the preferred winter roost site, as the birds settle not far from the hustle, the bustle, the traffic, and the lights. Once I saw them roost in a patch of pines flush up against the I-89 Interstate highway, where they were surrounded by the busy Burlington exit ramp.

One evening I watched them again as they came into the heart of the commercial district on Church Street. Round and round they flew in swirling clouds above the evening town crowd going to restaurants and theaters. It seemed as though they were looking for a place to land. I watched them fly over patches of trees at the edge of town that looked to me like ideal roosting sites, yet the birds still kept coming back into the center of town. Eventually they settled in several young cottonwoods next to Bove's restaurant. The birds were soon closely packed upon the branches, as ever more continued to stream in. That is when it suddenly occurred to me why the roost was here rather than elsewhere.

For decades there has been a heated debate about why birds join flocks. In the 1950s the British biologist V. C. Wynne-Edwards speculated that birds form communal roosts in order to assess their population size, so that they could then decide whether it was appropriate to reproduce, in order to maintain a stable population. To biologist's ears now this idea sounds about as plausible as that of the sun orbiting the earth to an astronomer. Wynne-Edwards's theory on bird flocking for the common good at the expense of the individual was termed "group selection." Animals can act cooperatively, but the specific example of Wynne-Edwards's theory sounded so ridiculous that his baby was soon summarily tossed out. But other ideas were spawned. William D. Hamilton came along and proposed the "selfish herd" hypothesis, which posited that animals form groups for their individual safety, using one another as shields. By being in a communal roost, the individual birds also profit from many eyes with which to see approaching danger. Hamilton's hypothesis made sense, and it also fit empirical observations.

The Israeli biologist Amotz Zahavi then vigorously promoted a third hypothesis that seemed plausible to account for other group behavior as well. Zahavi proposed that when birds roost communally they gain information on where to find food. His "information center," or IC, hypothesis provoked a good deal of controversy and generated hundreds of papers in the learned journals, perhaps because it looked suspiciously like a form of "group selection." Our data on raven communal roosts in Maine, principally the demonstration of naive birds being led to cattle carcasses after joining a roost

(Marzluff, Heinrich, and Marzluff 1996), gave perhaps the first empirically tested proof of Zahavi's IC hypothesis. (Of course, there are still those who, quibbling about mechanism, claim that the raven's behavior is "group selection," unless it can be defined in terms of tit-for-tat reciprocation among individual birds, not exploitation of individuals for information held within the group.) Neither hypothesis would explain the golden-crowned kinglets' tendency to congregate at night, because being in hiding and possibly also being in a torpid state, they could not instantly respond and escape a predator, even if one of the kinglets gave an alarm. Nor would they benefit from the group vis-à-vis finding food, as their food is highly scattered. What they do group for is warmth; unlike corvids, they seek body contact.

As I watched the crows in the city and wondered which if any of the existing aggregation hypotheses might apply, I was reminded of seeing other huge crow roosting aggregations elsewhere in both North America and Europe. Crows used to be thought of as strictly rural birds, but in the last fifty years they have started roosting in cities all over the world. There was clearly something significant in the way that the birds avoided the woods to be in the town lights and the bustle. It was not just a stray observation. The birds were free to roost in forests available within a half mile of the city, yet they had flown miles just to come here where they had to search long for a suitable landing site, eventually choosing the few available trees downtown.

Crows have greatly increased in numbers all over the North American continent over the last century, and they have been moving ever increasingly into cities. As one indication of both of these suppositions, I refer to a 1946 report in the *Oklahoma Game and Fish News* (Vol. 2: 4–7, 18), written by H. Gordon Hanson, a biologist of the Oklahoma Fish and Game Commission. His report gives the map locations of 47 "major" winter crow roosts in Oklahoma, those with

200,000 or more crows per roost. "It all started back in 1933," Hanson writes, "when it was first brought to the Oklahoma Game and Fish Commission's attention that the numbers of crows wintering in the state had begun to increase alarmingly and to spread over a much larger area." The crows fed on crops and came from the northern nesting areas in the prairie provinces of Canada where they "molest the nests of waterfowl in the duck factories." Although the winter crow roosts in Oklahoma had increased considerably in size and number, the Oklahoma Commission perfected metal cylinder bombs filled with steel shot and dynamite and then carried out an annual crow-bombing campaign. In eleven years government bombers bragged up a tally of approximately 3,763,000 crows killed. Given these data—and the probability that people living in cities would find the dynamiting objectionable—I speculate those bombed, and perhaps other, crow roosts were out in the country, instead of in cities as is so often the case now.

Crows are now federally protected birds, safe from the likes of state game commission dynamiters and others who took their lead to kill crows at every opportunity. Nevertheless, they also have their natural enemies, primarily great horned owls. Great horned owls nest in the winter, when they require much food to feed their fast-growing young. I have found two of their nests and heard them sing at night within two miles of downtown Burlington. The owls are perceived by crows to be their greatest enemies, and if they find one, the alarm goes out on the roost and dozens of crows converge quickly to harass the owl relentlessly, with the goal of driving it as far as possible away from where the crows will sleep at night. As my detailed observations of Bubo, a tame but free great horned owl, in *One Man's Owl* revealed, crows easily outmaneuver these great but somewhat clumsy predators in the daytime. At night it's a different story. It pays the crows to be in Hamilton's selfish herd at night, even as it pays them to be in Zahavi's information center at dawn. The two are not mutually exclusive. A communal roost can serve more than one function. The more benefits it confers, the more it is likely to evolve and to be maintained.

What I thought I was seeing in Burlington—crows trying to get into the city center—was no aberration. In one study of crow roosts in the Sacramento Valley city of Woodland, California, biologists Paul W. Gorenzel and Terrell P. Salmon document that the common crow's winter communal roosts are preferentially located in commercial (rather than residential) areas of cities, characterized by high nighttime light levels, and paved parking lots and commercial areas that have high noise and disturbance levels from vehicles and people. There are no great horned owls prowling downtown Woodland or Burlington or most other well-lit, noisy city centers.

That finally brings us back to the inexplicable anecdote about the crows "fighting" and "hacking" one another to death in a roost. What we see is not necessarily what is. The story is almost always in the details. The observer had mentioned that the episode occurred in the woods, hence it might have been in a location where one or a pair of great horned owls had access to the roost. It also could have been on a heavily overcast night in which much confusion ensued when the predators struck. With perhaps thousands of birds fluttering around in confusion, the predators would have maimed many crows, which would have then fallen, injured, to the ground. I have myself once seen an owl-killed crow, from which only a small portion of flesh was removed. When the observer arrived on the morning after an owl attack, he might have heard a bedlam of alarm from the injured crows. With many crows all around in the dark, the owls would not need to hold back. The carnage of crows that was described at the crow roost reminded me of a scene related to me by ornithologist Jeremy Hatch where a great horned owl raided a tern colony: "A dry pond-bed was strewn with about forty corpses: most headless, wings are generally torn off, or at least broken. Occasionally legs are gone. No evisceration. No plucking."

Crows in pain and in fear and not knowing what had hit them would blame and may strike out in frustration and anger at others near them. It is impossible to reconstruct what actually happened in the case of the warring crows, but there is one thing of which we can be absolutely sure. There are many advantages in different animals to aggregate in the winter besides keeping warm, but doing so in order to fight is not one of them.

A currently more public and more scary myth than that of the presumed warring crows is that the species might become extinct due to the much publicized West Nile virus. Crows occasionally die from this virus (and other causes), and there are cases of humans killed by it as well. After a dead virus-infected crow is found, "all" the crows may disappear in that area. But there is no reason to conclude that "the first thing that happens when the West Nile virus invades an area is that all the crows die—" ("The Silence of the Crows," in the *Washington Post*, 30 August, 2002), much less that we are "witnessing the disappearance of the crow from the American land-scape." I see no evidence that these birds are disappearing from "the American landscape"; there is no evidence that they are being depleted either by mutual warring or by the virus. If anyone finds a virus-killed crow, it is most likely to be simply where there are a lot of crows—a communal roost, and since roosts are strictly temporary, lasting only through the winter, it is therefore a given that the crows disappear after some die. We don't need another round of government bombing to kill 200,000 crows in a roost to get rid of a few sick ones.

By the end of July *summer is drawing to a* close. The autumnal symphony of cicadas and katydids has not yet started, but most birdsong has stopped. A hush comes upon the land. The insect songsters are still in their larval stages, and the young birds are out of their nests. Family ties are already broken and birds of many species then forge alliances with other juveniles, to form wandering bands along with their parents. Those vagabond bands include the huge populations of red-winged blackbirds and common grackles from the marshes. I meet them occasionally along the river-bottom cottonwoods and box elders. Flocks of tens of thousands of grackles, and sometimes of redwings, move through the late summer and fall woods like a giant steamroller, or perhaps functionally they are more like a giant vacuum cleaner. A vanguard will land ahead of the flock turning the freshly fallen leaves, looking for food, while others hurry along behind. The rear birds keep flying to the front to avoid the freshly searched ground and so they move forward, in a rolling action. Noisy black swarms of them occasionally loiter in the trees, joining and leaving the fray. And so they stay as flocks for most of the fall and winter as they travel ever farther south. They presumably migrate only as far as they need to find food. In early spring the blackbird crowds reappear. These birds disband from their flocks only for the short period of about three months after they return in early spring, and then some of them become semicolonial to nest.

While these birds leave and spend most of their time in flocks in more southerly regions, other species come to New England from their breeding grounds on the tundra and taiga of the Canadian shield where they live in pairs. They become gregarious and wander in flocks only as winter approaches. Following their food supply of seeds and of winter berries that vary hugely in kind and amount from year to year, they make only unpredictable annual appearances.

Most of the winter visitors from the north are finches—redpolls, pine siskins, evening grosbeaks, red crossbills, white-winged cross-bills, pine grosbeaks—that depend on tree seeds. But at least one fruit-eater, the Bohemian waxwing, and a grass- or weed-seed-eater, the snow bunting, also come in tight flocks from the north. Our resident goldfinches, which are solitary in the summer, also form their own winter flocks, but they stay. The purple finches stay only some-times, and they form loose, small flocks. The cedar waxwings that nest here in the summer also form winter flocks, separate from their close cousins, the Bohemian waxwings. Flocking up for winter is a common phenomenon. What accounts for it?

There is not likely just one explanation. As with communal roost-ing at night, there are instead many interrelated ones, and their rela-tive importance has been much debated in the scientific literature. The advantages of joining a flock probably include the "many eyes" effect of detecting danger, the previously mentioned "selfish herd" effect, to reduce predation risk as well as the learning effect of taking advantage of what others have experienced. Flocking specifically in the winter may reflect either dietary differences between summer and winter, or inability or reduced opportunity to wander in flocks while rearing young.

In the summer most northern birds must feed their young a high-protein diet so that they will grow quickly to adulthood. That means they must hunt for hidden and highly dispersed food, principally insects. In their hunting, individual initiative is at a premium. In the winter they can switch back to high-energy food, such as fruit or seeds, and many of these foods are in widely dispersed but large clumps that many pairs of eyes can locate more easily than one, where sharing costs little, and where there is risk to feeding alone when exposed in the open winter environment. But why do all the seed- and fruit-eaters segregate into their own species-specific flocks?

It probably relates to diet as well. Most of the seed-eaters are specialists for *specific* kinds of seeds, and by joining a flock they pool information of what is relevant to them specifically. For example, redpolls and goldfinches feed on birch seeds that are much too small for evening grosbeaks, so grosbeaks must forage separately. Snow buntings feed on the seeds of grasses, sedges, and other field plants. Evening grosbeaks have the strong thick bills with which to crack white ash seeds that the small finches lack. Crossbills have bills specifically adapted for prying apart the bracts of spruce and pine cones. Pine siskins have long thin bills suited for reaching the seeds under the bracts of hemlock cones. Given the different seed preferences and different tool kits needed to reach them, it pays for members of each species to join up and travel with its own kind. There are, though, glaring exceptions to the truism that "birds of a feather flock together." They are the resident birds of the winter woods that feed on insects. All over the world, insect-eating birds form conspicuous

A flock of goldfinches eating birch seeds.

Redpoll.

Evening grosbeak.

Pine siskin.

multispecies flocks. For about the last ten years I have kept a tally and made observations of these flocks in the winter woods in Maine because it seemed odd that these birds, which are solitary in the summer, would so dramatically change their behavior in the winter. Why would very different kinds of birds that feed on very different insects, and almost never on clumped seed or berry bushes, follow each other around in winter only?

Multispecies bird flocks are not unique to the winter woods of Maine. In the hot lowland forests of Tanzania in East Africa, I used to search for the noisy groups of a certain forest weaverbird, and having found a flock I would invariably see several species of bulbuls, barbets, and flycatchers traveling along. One theory for these groupings of insect-eating birds is that some birds of the group act as beaters that provide prey to another. For example, woodpeckers on the trunks of trees may chase off insects that fly off and that are then available to be captured by the specialists on *flying* insects such as flycatchers. It's the same idea as cattle egrets following buffalo to catch the insects they scare up, or some dragonflies following a large mammal through the grass, as happened to me in Botswana. (The dragonflies even chased me when I ran.) A second nonexclusive reason is safety in numbers. More eyes alert for predators such as snakes or hawks means less attention needs to be diverted to vigilance and more can be devoted to food-finding instead.

Like the bright-orange-and-black weaverbirds that I used to search for as markers of congregations of many interesting birds in Africa, so I look now for chickadees in the Maine winter woods. The other species commonly associated with a winter flock of chickadees near my cabin are two or three golden-crowned kinglets, a pair of red-breasted nuthatches, a pair of brown creepers, and sometimes also a pair of downy woodpeckers.

The chickadee flocks are probably the primary attractors for the other four bird species, because those species are almost never associated with each other in the absence of chickadees. I have never found creepers with woodpeckers, nuthatches with creepers, although all three are often alone. They either seek out and follow chickadees, or chickadees follow them, and the latter seems unlikely because a flock of chickadees cannot follow a half-dozen other species at the same time. Chickadees are also always by far the most numerous, the most noisy, and the most visually conspicuous members of any mixed-species flock. They make the most distinct "target." It is hardly possible to miss a chickadee flock, but it is hard to find the often quiet, unobtrusive, and hidden brown creepers, nuthatches, kinglets, or the downy woodpecker or two that may be with them.

The insect-eating winter birds minimize competition among themselves because each species forages on different trees, different parts of the same tree, or different prey. However, I'm doubtful that the beater effect, which may appropriately apply in the summer or in tropical forest flocks of presumably nonbreeding birds, would apply for these *winter* groups. Frozen insects are immobile and will not be chased off to become a target to a flock member. That leaves the many-eyes hypothesis for predator protection as a reasonable alternative.

I suspect flocking is advantageous at any time of the year, but it is constrained in the summer when the birds are tied to a nest site. In

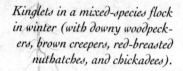

Kinglets in a mixed-species flock in winter (with downy woodpeckers, brown creepers, red-breasted nuthatches, and chickadees).

winter, when a limited food supply becomes a factor for survival, flocking may also become evermore advantageous, provided there is no competition for food. That is because it permits the birds to pay more constant attention to searching for food, and less on vigilance for predators.

In the Maine winter woods I almost always find the golden-crowned kinglets in groups of two to five individuals. Despite their small flocks, the group cohesion in kinglets is remarkable. The Austrian ornithologist Ellen Thaler, studying captive kinglets near

Innsbruck, found that the birds make special calls when approaching their sleeping place. These calls attract members of the troop of kinglets foraging together. A second assembly call draws the group into a cluster. Once together, the birds in the center of the cluster hunch their heads down into their shoulders with their bills pointing up. The birds at the edges tuck their heads back and to one side under their wing feathers. The group takes about twenty minutes to get into position in warm weather, but they bunch up in only five minutes when it is cold, although mated pairs and siblings always bunch up with each other in seconds. Apparently the kinglets recognize family members by voice and they are less inhibited to huddle with them than with strangers.

The average number of kinglets per winter troop near my cabin in Maine is probably too small for either the many-eyes or the selfish-herd hypothesis to apply. But if huddling is necessary for surviving cold nights, as can be deduced from physiological studies showing considerable energy savings of two birds huddling together, then a couple may be enough. However, body warmers are not likely to appear magically at dusk if each bird is moving at breakneck speed all day through the dense woods, looking for the nearly invisible caterpillars it likes to eat. Attracting and keeping vocal contact with others throughout the day may be a key component of their winter survival, especially in dense coniferous woods where kinglets are not only rare but almost invisible. The availability of body warmers at dusk cannot be left to chance; losing only one or more members per troop might doom the rest to freezing to death on some cold nights, especially after a day of poor foraging. If so, then it is no surprise that no kinglet winter flock, even a tiny one, is ever silent for more than several seconds at a time. The birds try to keep in contact. If they should get separated from a traveling companion, then it's a safe bet that by finding a noisy chickadee flock, they would soon meet up with another of their kind. That, at least, is how I reliably find kinglets.

Wondering *what the crows roosting in town eat*
in the winter, I collected bagfuls of their regurgitated pellets under the roost and picked the pellets
apart. These undigested remains of what they had
eaten that didn't pass through their digestive tract
contained mostly berry seeds. Some pellets consisted primarily of wild grape seeds. Others were
masses of seeds from the common *Viburnum*
species, and some contained wild holly seeds. In
early spring many were almost pure bundles of
undigested staghorn sumac seeds. Almost every
kind of winter berry was represented in these crow
pellets that can be as much fun to pick apart as owl
pellets full of bones and fur.

Only nine of the thirty-eight local species of
berries that I know of ripen and rot quickly. These
are the summer berries, such as strawberries, June
or serviceberries (*Amelanchier*), raspberries, black

berries, blueberries, and chokecherries. They ripen in a progression from May through August and they spoil within several days. That leaves twenty-nine winter berries, which all ripen in the early to late fall. They are not sweet, but they last on the branch through the winter.

I have come to notice these winter berries as a consequence of watching birds. Berries and birds are intertwined in an ancient and complex mutual relationship that is as intricate and interesting as that of flowers and bees, although it is not always as visible or obvious because it proceeds over time spans measured by seasons rather than minutes.

Watching the different kinds of berries that are specific to winter can be a slow sort of sport, unless one has a good view and keeps a long-running score. I am lucky to have a beaver bog nearby which I routinely check to see what's happening. Many berry species can be found there. The beaver pond is surrounded by a virtual hedge of arrowwood (*Viburnum dentatum*), a species whose dark-blue berries show up conspicuously because they are attached to the twigs by bright yellow stems.

Arrowwood berries provide a considerable bird feast in the fall. During the autumn of 2000, I counted and weighed the berries from twenty arrowwood bushes, estimated the number of bushes surrounding the pond, and calculated that the little beaver bog yielded slightly over a ton of bird food. From early September and into November, wave after wave of hundreds of robins came through, feeding for several days before moving on. Bluebirds, blue jays, starlings, grouse, and pileated woodpeckers also fed on these arrowwood berries, and I found not one berry remaining by early March, when the resident winter birds would presumably need them most. Indeed, by February arrowwood berries are dry, their yellow attachments to the twigs are brittle and faded, and the remaining berries then drop off. Thus, although fed on by many birds, these berries are apparently adapted for *fall* migrants.

Surrounding the beaver bog, though in much less abundance, are also maple-leaved arrowwood (*Viburnum acerifolium*) and nanny-berry (*Viburnum lentago*). Both species have blue to black raisinlike berries, but unlike the *V. dentatum*, these stay attached by stalks that do not turn brittle. Furthermore, these berries are not consumed by the fall migrants. They can stay on the bushes all winter and they are fodder for resident winter birds such as grouse, purple finches, blue jays, and crows. If not fully removed, they then also feed spring migrants.

Winterberry, the common name for a holly (*Ilex verticillata*), should presumably also feed winter birds. It grows in swamps and wet places, bearing crimson berries. A bush laden with these berries, after the leaves are dropped in the fall, shines like a red torch that signals birds to come eat and ultimately to disperse the seeds. Yet, I had routinely seen resident winter birds, such as woodpeckers, chickadees, kinglets, blue jays, and grouse, in and around these bushes, and the berries stayed for months, untouched.

One early November day a few years ago, I saw a flock of about two dozen migrating robins descend on a large holly bush covered in thousands of berries, and when they flew on they had not left a single berry uneaten. Having been alerted, I now routinely see robins (and no other birds) eat these berries in the fall. The few berries that don't get taken in the fall remain until about February, when their color fades to yellow-brown and they start to drop off. Thus, even the quintessential "winterberry" caters to a select clientele of fall migrants and not to the resident winter birds at all.

One of the most enigmatic berries to me is the highbush "cranberry" (*Viburnum opulus*). This berry is bright red, like those of holly in late fall, but the *V. opulus* berries decorate bushes through winter and well into early spring, which is why they are often planted as ornamentals. The bright red is a conspicuous enough signal to birds, yet year in, year out the berries stay on the bushes for months. An

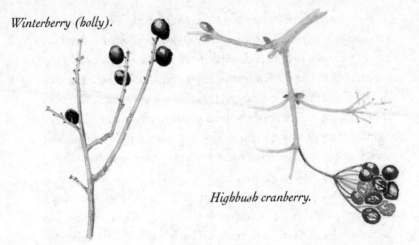

Winterberry (holly).

Highbush cranberry.

abundance of highbush cranberries grow among the other berries in a winter berry patch I have cultivated next to our house in Vermont, and while birds have routinely feasted on arrowwood and nanny-berry, they have always left the highbush cranberries alone. They are not poisonous—they are depicted as suitable fare in recipe books on wild fruit. Taken raw, they taste sour, but when boiled and sugared they are not objectionable, at least to humans. I was perplexed that such a "typical" bird berry was not eaten by any of the fall migrants nor by any of the other birds that were around it every day, for months. Surely, some bird must find them palatable.

Then on February 23, 2000, the mystery was solved. A flock of eighty to one hundred Bohemian waxwings (only sporadically seen here in winter) arrived and landed on a sugar maple tree above the large highbush cranberry bush at the edge of the field near the house. One bird dove down to the berries, several flock members followed, and then the whole flock descended en masse. Within a half hour the large clump of bushes that had been heavily laden with berries was stripped totally bare.

Flocks of both Bohemian and cedar waxwings were locally common that winter. Both species of waxwings were also feasting,

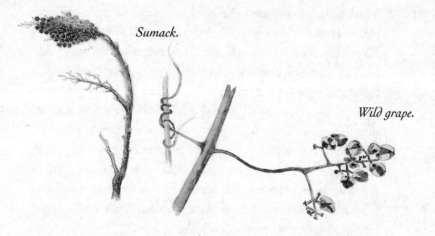

Sumack.

Wild grape.

separately, on the berries of ornamental hawthorn on the University of Vermont campus. Both waxwings live entirely from berries in the winter, but unlike most other birds, their summer diet is frugivorous as well. Thus, the highbush cranberry is perhaps the choice of the discriminating berry-bush specialists, and highbush cranberries remain bright and juicy throughout the winter as though they were being preserved specifically for late winter or early spring migrants.

In the following summer none of the *Viburnum* bushes flowered and thus no berries were produced in the fall. However, I again saw flocks of Bohemian waxwings in February. This time I found a large flock in brushy young woods. I was surprised to see them there, because I saw no berries. The birds seemed to be picking buds, or so I at first presumed. Looking closer I suddenly discovered that they were picking sparse, hard-to-see, dry, shriveled, black-blue buckthorn (*Ramnis cathartica*) berries. Therefore, I did not find the berries by myself. I found them by joining the flock, and birds probably find berries similarly.

Staghorn sumac is also available all winter long. It has tightly packed small dry fruit that have no visible meat. The massed fruit are almost bare seeds covered with a hairy fuzz as though designed to be unpalatable. Like highbush cranberry, this sumac ripens in late summer and remains uneaten on the bush almost all winter. The berries are preserved not only by acid, as is highbush cranberry, but by dryness. Yet, in early spring, when few other berries are left and choice is correspondingly limited, they are finally eaten by robins, starlings, crows, and a variety of returning migrants including flickers, thrushes, and catbirds. Apparently sumac saves energy by not offering a fleshy fruit with sugar, fats, and protein, and it gets away with that stinginess both proximately and ultimately because the fruit's very long branch life extends beyond that of the more perishable competition. (Sumac fruit have, to human palates, a lemony taste and are a good source of vitamin C, if boiled in a tea.)

Many birds require berry crops for fattening up on their long fall migratory flights, putting on as much as 10 percent body weight per day. That impressive fattening feat involves adjustments of gut length and other digestive adaptations for berries that allow for rapid food processing (Karasov). The fruit's nutritional content depends on the season for which their dispersal is tailored. Thus although the highest-quality (highest energy content) fruits contain fat and sugars, that food (especially fat) causes rapid fruit spoilage due to microbes (Stiles). Low fat and sugar contents, as well as high acidity and low water content all help to prolong branch life, with staghorn sumac being the extreme of that strategy. Its fruit can be retained without spoilage for about eight months after being produced. Extreme? Well, maybe not entirely. The tomatoes we get at the supermarket may be a close analogy. They are selected for long-distance travel from California and for long shelf life, unlike the garden variety we grow for good taste. As with wild fruit, the nutrients that make them taste good also cause their rapid spoilage, and our commercial varieties of fruit are selected, like many winter fruits, for longevity.

If the berry is adapted to the bird, then it stands to reason that the bird is adapted to the berry. But aside from specific digestive physiology, there is also the perhaps even greater problem for the bird of locating the often widely dispersed berry bushes.

Seeing robins, bluebirds, starlings, crows, and waxwings descend on their berry bonanzas suggests why these winter frugivores fly in large flocks. The "many eyes" hypothesis posits that animals in groups can more easily spot danger. Might it also allow them to more easily locate the scattered food bonanzas? The latter has a cost, namely that any one individual who finds a berry clump has to share it. However, that cost is not great if the clump is ample and the flock has to keep moving. So, joining up with a hundred pair of other eyes can pay off in more ways than one, especially if everyone is in a rush and not interested in staying around.

Many migrants fly to staging areas in between their migratory endpoints where they have previously found food and where they fatten up for the long journeys ahead. After seeing the berry-gulping flocks on their migrations, I started to cultivate winter berry bushes despite the fact that most don't feed the resident winter birds. It pleases me to think that after the bears have fattened up on the blueberries, chokecherries, and wild apples, there are still plenty of winter berries to go around, providing fuel for not only crows, but also the long-distance wanderers on their stopover from the north.

While searching for kinglets in the Maine winter woods during mid-December 2000, my students and I found the fresh spoor of a black bear on new snow. We had never seen a bear track so late in the year. When there is little food, the bears den up as early as mid-October. It had, however, been a fall with a heavy crop of both beech-nuts and acorns. Wild apples had also been abundant on the old overgrown farms in the surrounding hills.

The bear had passed by only hours before, and we took up the chase, hoping to find its wintering lair. The bear had traveled without stopping to rest, walking past a calf carcass that we had laid out to feed ravens. Normally such prime veal would attract a bear miles away and then cause it to gorge. But the fact that this bear did not stop to feed prior to its long winter fast was not surprising. By

December it had probably stopped looking for food and was now searching for a denning place.

The bear started to switch from prehibernation to *hibernation* physiology. This switch is triggered by chemical signals in the blood. Once in hibernation, bears will stay in their winter dens for as many as five months at a time, conserving their hard-won energy resources (fat) that they have accumulated during their feeding frenzies of the previous fall. Running around searching for food when little is to be found is deficit energy spending. Evolutionary logic dictates that appetite would be suppressed in a bear ready to hibernate, because a hungry bear would continue to be a food-searching bear, and deficit-spending bears become dead bears.

There are two ways to try to beat an energy crunch brought on by winter. One approach, used by the kinglet and human beings alike, is to work harder and harder to try to maintain a profit margin, even as we pay ever-higher heating costs in the face of ever-dwindling resources. There comes a point, however, when it is better to drop out in an effort to save as much energy as possible. The latter is sometimes a matter of necessity and it is common in many animals in winter. It may even occur in humans, given the right circumstances. Here in midwinter at the high latitudes of Vermont and Maine, I start to feel sleepy at about 5 P.M. and I have little trouble curling up in a snug bed as soon as it gets dark. In the summer at that time there are still four more hours of daylight to come, and I would still be running around. Of course my semi-hibernation tendencies are blunted ever so slightly by social pressures. In my culture it is just plain lazy to sleep fourteen hours per day. So, as a result of social conditioning I routinely extend my winter day with artificial light in the evening and with caffeine in the morning. Most important, my natural tendencies may be suppressed because I'm not on a stringent diet that winter would normally impose. My calorie intake is undiminished and sufficient to keep up my energy level. Unlike the bears in the woods, we New Englanders don't need to go into hibernation in October when the nuts run out.

My students and I did not succeed in tracking our bear to its den. If we had, we might have found it under a brush pile, in a hollow tree, under roots, or upon a heap of branches in a stand of dense young spruces or balsam firs. Bears may even curl up and hibernate in the open with the understanding that they eventually will be buried with snow. Grizzlies dig their dens, moving up to a ton of earth to carve a tunnel into a hillside, their snug hibernation chamber at the end of it. They cover the floor of their chamber with bedding material of branches, grass, or duff scraped from the ground nearby.

Bears may prepare their overwintering den a few days or weeks before entering them. Some wait for a big snowstorm before finally crawling into their prepared sanctuary. They are flexible and individual differences abound. However, most bears engage in a feeding frenzy in late summer and early fall, in which they down about five times their normal food intake, putting on a five-inch layer of fat. By late fall they slowly lose their appetites until they eat nothing and when they leave their dens in the spring, they are no more hungry than before entering. Instead, provided they retain some fat, it takes them a long time to regain their appetite. Appetite suppression during hibernation is probably under the control of leptin, a "satiety" hormone secreted by fat cells that circulates in the blood and affects the appetite centers in the brain (Ormseth et al. 1996). In spring, leptin levels decrease and appetite increases.

When are bears in hibernation, if ever? This common question is not a good scientific question, because the best answer is "it depends." We often seek precision by pigeonholing through definitions, whether with respect to what is right or wrong, alive or not alive, hibernating or nonhibernating. The problem is that animals don't stay within such simple boundaries. They don't obey rules, so

Three-month-old nonhibernating cub climbing on hibernating mother.

any precision that might be gained artificially through a definition is apt to slip away at any moment in any case in point.

Overwintering bears have many physiological and behavioral characteristics, but they were for a long time not considered to hibernate, simply because their body temperature showed only modest drops and hibernation was defined in terms of low body temperature. A bear's body temperature during hibernation remains near 35°C, only slightly lower than the 37° to 38°C or so when active. At a body temperature of 35°C a bear may be slightly sluggish, but it is by no means unresponsive to disturbance, especially from human researchers who would dare to enter its lair to take its temperature with a rectal thermometer, or stick it with a syringe to draw its blood in trying to track down the marvels of the bear's hibernation physiology. It turns out, however, that the key to a bear's hibernation is not to be found in the temperature of its rectum. Instead, diagnostic characteristics are discerned through the bear's appetite physiology, waste metabolism, water balance, and bone retention despite lack of exercise. Indeed, the marvels of hibernation concern many medical matters of acute practical relevance to humans, especially as regards aging, space flight, and osteoporosis.

One of the first issues of hibernating bears to be studied was how, despite maintaining a high metabolic rate (high body temperature) the bear still does not need to drink or urinate all winter. We can, like bears, also go without food for a long time, provided we have body fat. But we can't get along without water. If we were to spend considerable time in a bear's den in winter we would, even without sweating, quickly dehydrate due to urination. But if we shut off our kidneys, then our metabolic waste, principally urea, would pile up in our blood until it poisoned us. Urea is our vehicle for getting rid of nitrogen, which becomes a waste product after we digest protein or nucleic acids. A bear does not urinate all winter. Thus the question is: Does urea not poison the bears or don't they produce urea? To find out, physicians Ralph A. Nelson and Dianne L. Steiger from the Carle Foundation Hospital at the University of Illinois teamed up with game biologist Thomas I. Beck from the Division of Wildlife in Colorado to try to examine the urea content in the blood of hibernating bears. But how to get the blood? Bears in their winter dens are alert enough to be intolerant of people with hypodermic syringes. In part to get more compliant subjects for their project, the researchers did the next most difficult thing, they live-trapped bears in the fall and equipped them with radio transmitters that could be used to track down their subjects later when they were denned up. There, the bears were tranquilized with chemicals (Rompun) from a dart gun that made then more tractable and the task of taking their blood easier. A total of 76 blood samples from 48 bears were collected and analyzed.

Since hibernating bears metabolize mostly fat, they do not accumulate huge amounts of urea in their blood. What small amounts that they do produce they convert into creatine, which is nontoxic. Additionally, instead of becoming a toxic waste, the nitrogen wastes in hibernating bears are biochemically recycled back into protein; hence no loss of muscle mass is experienced even as they don't exercise. Thus a hibernating bear never needs to get up to take a drink or go take a leak all winter. Water is conserved because none is needed to flush out toxic wastes, and the animals stay in shape. But that alone does not make the bear the ultimate enviable couch potato that it is. Other physiological wonders of its fitness continue to be elucidated.

We require mechanical stress of exercise on our skeleton to maintain bone structure and function, as was dramatically illustrated during weightlessness experience in space (Johnson 1998). Bone mineral content loss, depending on specific bones, was 3.4 percent to 13 percent, even on a 17-day space mission. Muscle volume decreased at similar rates, raising concerns for the effect of extended space travel on astronauts' health (White and Avener 2001). Paraplegic accident victims suffer similar bone-density loss, with 30 percent lower-body bone mass depletion within six months of losing the use of their legs. Neither space nor weightlessness as such is therefore the real cause of debilitation. The real culprit of osteoporosis and muscle mass loss is physical inactivity, and to counter these effects, the Soviet space program, which included a record-setting mission of 366 days, emphasized intensive exercise. Bears do without the exercise and suffer no ill effects. In their long evolutionary history, those that could not tolerate the rigors of prolonged inactivity were weeded out.

Scientists at the NASA Ames Research Center have studied the effects of inactivity in healthy young volunteers who got well paid to lie flat on their backs in bed for a few days to a few weeks at a time (Miller 1995). Since 1971 more than 500 participants at the Ames Center have proven the dramatic implications of sedentary lifestyle to humans. There is not only bone loss and weakened muscles, but also slower absorption from the gut and prediabetic resistance to insulin. The scientists concluded that the physical stresses placed on the body during space flight are virtually identical to those of prolonged bed rest, or those of a hibernating bear. Again the question was: How does the bear's body stave off bone loss?

Most of the American population subjects itself to the physical stress of inactivity. One in three Americans over fifty is completely sedentary. Therefore, our muscles, deprived of exercise, become resistant to insulin that normally promotes the absorption of glucose; blood sugar reaches dangerously high levels whenever we consume sugar-containing products and so we risk the onset of adult (Type II) diabetes. Our bodies are not adapted to inactivity. In our evolutionary history, in contrast to bears, exercise was a constant, and we're not made to tolerate being idle for long. We're adapted as long-distance endurance predators, as I've elaborated on in *Why We Run: A Natural History*. Inactivity adversely affects every organ system in the body, at least so long as we continue to eat. However, I suspect that a caloric surplus could be a relevant variable, too, since that is often the result of inactivity.

Ralf Paffenbarger, a physician at Stanford University, studied the effects of the lifestyles of 17,000 Harvard students for twenty-five years after they graduated from college and concluded that exercise is a prime variable for health and longevity. That is, the stresses of inactivity mimic the aging response. Every hour of vigorous exercise as an adult was repaid with two hours of additional life span. There is, obviously, a limit to the benefits of human exercise or else more exercise could make us immortal. Instead, too much exercising increases the aging process as well. I suspect the debate of optimum exercise for maximum longevity may relate less to how much exercise we get than to how many calories we take in versus how many we burn off. There is a correlation between eating less and having a longer lifespan. But of course starvation shortens life, and there is thus also a correlation between eating more and having a longer lifespan. The difference is in the *range* of food intake versus the amount of exercise.

Rest does not as a rule speed up the deteriorating effects of aging, since hibernators that enter torpor have a longer life span than nonhibernators. That is not surprising, because during hibernation physiological functions are put on hold, presumably those that result in degradation as well as those of regeneration. Thus, the long period of low body temperatures characteristic of hibernation are like death to the animal, so that in effect its life span may be extended, even if the number of hours it spends *living* (as we define eating, defecating, moving, and sleeping) are curtailed. (It would be of interest to know if, subtracting the hibernation time, hibernators have the same or different life spans than their nonhibernating congeners.)

Hibernating bears accomplish metabolic feats that, if we knew their secrets, would likely lead to cures for many human ills. They have the secrets of *how to* survive lack of exercise, and then, after five months of resting, of how to get up and walk up a mountain. In all of those months of what amounts to bed rest, they suffer no bed sores. They have marginal loss of muscle mass and no change in muscle fiber type. Despite their non-weight-bearing position for months at a time, they do not suffer from bone loss or osteoporosis. After burning fat for fuel for months during which their cholesterol levels become double those of humans and those they have in the summer, yet they still don't suffer from hardening of the arteries or gallstones, conditions resulting from high cholesterol levels in us. Most of the enigmas that have been revealed in hibernating bears have not been solved, maybe because bears just can't be studied as conveniently as lab rats. We can be reasonably certain, however, that once we understand how bears hibernate through the winter, we will also have a larger window into ourselves. We inadvertently simulate a hibernation-like state of inactivity in our modern environment, a new state of nature to which we are not well adapted.

The vegetation *in northern New England is* at its lush green peak by mid-August. Yet, the azure blue butterflies have not been around since mid-May, and the tiger swallowtails since mid-June. Each species appears and dies in its specific time slot. The pupae of most insects have by now been arrested in their development for a month or two, and they won't revive from hibernation to develop into adults until their specific times next spring or summer. Meanwhile, the monarchs have finally arrived from the south, and some of their fast-growing caterpillars are already developing into pupae that a week later turn into adults that will begin their journey back to the south when their food, the milkweed plants, begin to dry up. Ants are still tending the aphids that they milk for their honeydew, but soon they will begin to bring them to safe underground quarters for the winter. Bears and woodchucks are fattening themselves up. Cicadas call shrilly during the day and the constant chirping of crickets, grasshoppers, and katydids continues night and day. But in two more months these songsters will be stone-cold dead, perhaps even before the first fall frosts. There are signs all around of profound change about to happen, as the necessary physiological and behavioral adjustments for the coming winter are being made. In chipmunks, certain birds, honeybees, and us, the most important preparation for winter is storage of food.

In our garden the apples, pumpkins, and squash are still ripening, but the onions, garlic, carrots, potatoes, and string beans are ready for harvest. We're starting to freeze, can, and dry the bounty for winter, and I've been sawing wood and stacking it in the cellar. The farmers' barns have long since been full of hay to tide the cows over through May, and the unmowed fields and orchards are ablaze with the yellow bloom of goldenrod and the blue New England asters that are abuzz with honeybees topping off their own fuel depots in their hives.

For my family these rituals of food and wood storage fulfill some basic urge, but they are certainly not obligatory, thanks to our twenty-first-century transportation and monetary systems. However, just one hundred years ago, the work of food gathering and storage was a necessity for survival for those who did not rely solely on hunting.

Among mammals, food hoarding can be an alternative to hibernating or migrating, but only a small percentage of the total number of species on the planet store food. And while we may be able to do it on the grandest scale, humans are by no means the most spectacular of food hoarders. In late summer, the pikas (*Ochotona princeps*), a small relative of the hare that live in the high mountains of the West, collect grass, dry it in the sun, and then pack the hay into dry cavities under rocks for winter food. After the leaves turn and fall in the autumn, beavers all over North America begin felling trees and saplings, and dragging them into the water to make huge underwater food caches near their lodges. The icy water keeps the bark fresh throughout the winter, and the beavers live by feeding on it for about six months. Some squirrels and many other rodents, including deer mice, pocket mice, kangaroo rats, and hamsters, stockpile seeds that reduce or eliminate their need for torpor.

A frozen apple on a white birch twig cached there by a red squirrel.

Among birds, long-term food storage occurs with generally northern species (Källender and Smith 1990), and in those species that exhibit no or only modest nocturnal torpor of several degrees. Food-caching behavior is found almost prominently within two families; some of the Paridae (chickadees and kin) store food for winter, and most of the Corvidae (crows, jays, magpies, nutcrackers, and ravens) do so. There is much variation within any one group. At one end of the spectrum of behavior are chickadees, which when encountering a food bonanza that they cannot eat all at once will store some of the food, stuffing it into clefts and crevices, and come back later for it. However, there is no evidence that they lay up food stockpiles for long-term use. On the other hand, European marsh tits and nuthatches may depend on stored beechnuts for a significant part of their winter diet.

Within the crow family there is also a gradation of behavior, ranging from temporary storage of a surplus to long-term storage that sustains the animals through the winter and well into the breeding season. Like pocket mice, kangaroo rats, hamsters, and chipmunks that are adapted to carry off surplus food in their two expandable cheek pouches, corvids that cache have an expandable throat-pouch under their tongue for carrying food to storage.

The super-cachers among the Corvidae are probably the nut-crackers, the Eurasian, and the Clark's of the mountainous regions of western North America. The first stores mainly hazelnuts. The second lives primarily on the seeds of several species of pines, such as pinyon, limber, and whitebark, which ripen in the fall (Vander Wall and Hutchins 1983). A single nutcracker routinely collects tens of thousands (and as many as 30,000) pine seeds and stores them in 2,500 individual caches, commonly in windswept south-facing rock ledges that may be fifteen kilometers from where the seeds were picked. Months later the birds remember the location of about 80 percent of these caches and come back to retrieve the seeds. The nutcrackers' ability to live off the seasonal seed crop (Vander Wall and Balda 1981) depends on their astounding memory (Vander Wall and Balda 1983) for specific sites that humans would likely find impossible to match. The seed caches that are not retrieved are, in the long term, not wasted because they propogate the food source on which these birds depend.

Common ravens, when feeding on carcasses in winter, cache meat. However, their food stockpiles last only relatively short periods of time, because the caches get buried by frequent snowfalls. Additionally, meat is perishable or is easily found and dug up by carnivores with a keen sense of smell such as shrews, coyotes, and foxes. Thus, their caches are more for immediate, not long-term use.

We have only a glimpse into the general pattern of how ravens maintain an energy balance in winter, but that peek is intriguing. Ravens begin to nest in late winter, and all of the young are kicked out of the parent's territory by late summer, if they have not left on their own already (Heinrich 1999). The young then follow the food—preferably a constant supply of fresh food such as that provided by a pack of wolves (Stahler, Heinrich, and Smith 2002). In areas of the north woods where there are no longer wolves, the ravens hunt small prey and also scavenge from carcasses opened by other carnivores. Adult pairs stay year-round in their territories and defend carcasses they find there. Wandering juveniles are excluded from the feasts, unless they band together to overpower the resident adults. They "gather the troops" by joining communal roosts at night, and naive hungry birds can get food by tapping into information of food resources, by following those that take the initiative to leave predawn to fly to a food bonanza such as a deer or moose carcass that they had discovered or fed from previously. The problem is, there is a lot of competition at the carcass, and not all of the crowd of dozens to a hundred or more can be assured continuous access to the few available feeding spots. But there is a solution: Those dominant birds that can get at exposed meat haul off as much as they can and hide it, to retrieve later when the gang may have removed most of the meat. And those that couldn't get through the crowd? They too have a solution: They closely observe where the dominant birds hide "their" meat, wait until these cachers are out of sight, and then recover the other's cache either to eat the food right there or hide it elsewhere. In turn, a bird trying to hide food avoids potential cache-raiders and positions itself to be out of sight from them. In the intense social interactions among the birds at carcasses in the winter, those that can best anticipate the intentions of competitors (and the potentially dangerous carnivores also at the carcass) are most likely to be reliably fed. Such a scenario, where one or a few individuals may try to control a resource in a crowd, is currently a prime scenario for consciousness or what is known among animal behaviorists and biologists as "theory of mind." This is a much different sort of mental facility than memory, and ravens, although possibly the most intelligent of birds, do not exhibit impressive long-term memory for cache locations. Cache locations are remembered for two weeks, and a month may be their limit (Heinrich 1999). But given the ravens' lifestyle and their food, that's probably sufficient.

Each animal's lifestyle has its own unique opportunities, and requirements and constraints. That of the gray jay (*Perisoreus canadensis*) offers intriguing contrasts to that of the raven. These fluffy, diminutive members of the crow family look to me like oversized chickadees. They are everything a crow or a raven is not: tame to the point of seeming friendly, vocally muted, restrained. They are silent fliers who glide on mothlike wings. They never aggregate in big crowds, generally don't feed on carcasses, and have a soft look and none of the raven's or the crow's bold sharpness of eye. Instead of flying from you on rapid wingbeats ripping the air, gray jays more typically silently glide up to you. Gray jays, formerly called Canada jays, are the north woodsman's endearingly named whiskey jacks, and camp robbers, who are always looking for a handout. That is why they are so intriguing; they seem to have no visible means of support. Yet, they live in what appear to be barren spruce forests, far from human handouts. They live only several miles from my cabin in Maine, and I've met them in willow thickets on the Noatak River on the North Slope in Alaska. They are one of the very few birds that survive year-round in the northernmost taiga, breeding as early as March, often two months before the snow has melted. How do they manage it? The answer to this bird's riddle probably has less to do with either superior memory or intelligence and a lot more to do with their saliva.

William Barnard, an ornithologist from Norwich University who studies a small population of these birds in Vermont's Victory Bog, tells me that their saliva is "amazing stuff." Most spit is designed to ease the food down the esophagus. It must be slippery and non-sticky. This birds' saliva coagulates on contact with air, to become viscous and sticky. In short, once extruded, it becomes glue. It is a very important glue to gray jays. These birds live in environments with deep snows all winter long, where food caches on the ground or in the snow can become unavailable from one day to the next. Gray jays are unique among Corvidae for routinely storing food *above* ground. And that's why their spittle is important. It's the glue they use to cement their food caches to trees, and that capacity allows them to forage in late summer and fall and safely store food for winter. It keeps their food away from numerous ground marauders, and at the same time alleviates the necessity of digging through feet of snow.

Gray jays begin nesting in March, about two months earlier than their cousins, the blue jays. At that time in late winter in the Northern Hemisphere, and on high mountains, they can still expect numerous snowstorms and days of subzero temperatures, if not weeks of subzero nights. So aside from sticky saliva, the next critical component of their energy strategy is nest construction. Unlike the blue jay's flimsy see-through nests of bare twigs and rootlets, those of the gray jay are bulky, deep, and well-insulated cups lined with fur and feathers that cradle and keep warm the clutch of three or four grayish, olive-brown-spotted eggs.

The early nesting by the jays must have an advantage. We don't know for sure what it is, but a study by Dan Strickland in Quebec and Ontario provides clues. Gray jays may seem like very friendly birds because they readily approach humans, and Strickland (1991) found their nests by offering the birds prized nesting materials, primarily cotton, facial tissue, and grouse feathers, and then following them. From a study of 470 color-banded nestlings in 179 nests he found a surprising social structure centering on food caching.

In Quebec, food caching for winter by a resident pair of gray jays starts in late summer when food is most abundant, so caching then makes sense. With plenty of food available and little or no competition for food, most corvids are then tolerant of each other, especially toward family members. But not gray jays. Strickland was surprised

Gray jay "gluing" food onto tree.

to see constant aggression and chasing of the juveniles within the family group. The result of that intrafamily strife was almost invariably that only the most dominant of the brood remained in the parents' territory. (Those that left sometimes joined up with other pairs whose breeding attempts had failed.) It was not clear why the parents should tolerate a freeloading offspring all winter, but some evidence (Waite and Strickland 1997) suggests that the lone stayer eventually pays its due by helping at the nest in the parents' next nesting attempt. Besides, parenting is *always* costly. And parents have to do whatever is necessary for the offspring's survival. But why would a young bird fight with its siblings almost to the death to be able to stay with its parents?

To survive the coming winter, the young need to store up food. However, young corvids, like the young of most other birds, require experience to become good foragers and cachers of food; especially those that like gray jays and ravens may learn to forage for rather "exotic" fare, such as the blood-engorged ticks that ingest moose blood in winter (Addison, Strickland, and Fraser 1989). Possibly, young gray jays are too inexperienced and cannot quite find and lay up enough food to survive the winter, and so their only chance of surviving is if they can rely on a partial winter subsidy from their more-skilled parents. If so, then the parents are obliged to provide it, or they lose their genetic investment.

The parents, however, are also limited by food and they can't shortchange future reproduction. They can perhaps support one freeloading youngster through the winter, but not three to four survivors of a full clutch of eggs. If all of the clutch survived, if all of the young of any one clutch stayed, then the *whole family* could starve. Thus if brood reduction in the winter is inevitable, then it's better to force a fight and make it happen in advance of the food crunch. The evicted subordinates' advantage lies in leaving while they still have a chance of finding a pair of adults whose nesting attempt failed, that would be less predisposed to evict a starving, persistent juvenile. That is, the evicted young in effect parasitize the parental instincts of failed breeders.

I contrast now the behavior of the gray jay with another animal that also survives the winter only because of the energy resources it collects during the summer months. This animal's social system is also crucial for making the storage of food energy possible. However, there is one large difference between it and the gray jay. In this animal only the mother survives the winter, along with her tens of thousands of daughters. The male offspring, who are unable to feed themselves, get kicked out of the warm nest or are starved in the fall by not being regularly fed. In this case, unlike with the gray jays, all the daughters participate in helping the mother rear subsequent broods of offspring. What I'm talking about, of course, are honeybees (*Apis mellifera*).

Honeybees manufacture a special product with a high energy content that has a nearly indefinite shelf life in the well-regulated microclimate of their nest. The raw material for honey, nectar from flowers, is carried in a unique distensible stomach that serves as a large bucket. The pollen or "bee bread" used to feed the young is packed into a special hair-structure on each hind leg. The bees make wax, and use it to build finely crafted receptacles to the most precise specifications to store the honey and the pollen, but separately. A colony of honeybees may routinely store a hundred pounds of honey (plus numerous pounds of pollen) for the winter. Special climate-control mechanisms are employed to maintain the wax receptacles holding the honey and/or pollen at precisely the right temperature, to keep them from melting while also soft and malleable enough for shaping. Ventilation and proper humidity control are used to concentrate the honey and control molds. Special orientation mechanisms involving a superb internal time sense allow the bees to use the sun as a reference point to navigate quickly and efficiently between foraging and the food storage sites at the nest. As a result of their superb energy-storage mechanisms, honeybees are the only insects in the north that don't hibernate and that maintain a high body temperature all winter long. Bumblebees, which also collect honey and pollen and which individually work longer and faster than honeybees, do not lay up food stores for winter. There is no need. They only store for a rainy day, converting their wealth directly into offspring, then the colony disintegrates in the fall and only the fertilized females (new queens) stay alive to hibernate in cold storage in the ground.

As I sit here at my house in Vermont, next to my four beehives in mid-August, the goldenrod in the neighbor's fields is in full bloom and my bees are busily harvesting nectar from it to make honey to fuel their energy metabolism through the winter. Pollen is collected often simultaneously with nectar, and this pollen will be used in early March when the queen starts to lay eggs. It will feed the larvae that grow long before foragers can again leave the hive to hunt for food. Goldenrod is, next to asters, the bees' last major crop of the year. Some late wild aster will top off their winter larder, but after that the bees here in northern New England will have to wait almost a half year before they'll see another flower.

It is during this intervening time, when the honeybees are confined to their hives, or nests, and the bumblebees are hiber-

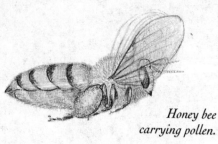

Honey bee carrying pollen.

nating in torpor underground, that the kinglets, not much bigger than big bumblebee queens, are busily foraging in the spruce thickets. As far as we know, they do not rely on or use any food larder. Kinglets have never been observed to cache food. This does not, however, totally exclude the possibility that they do it. Maybe they've not been observed under the right circumstances. Potentially, kinglets are at least as intelligent and programmable to store food as are bees. But they don't live in a hive appropriate for making and storing honey. Kinglets diversify their options by being partial migrants, and animals that may have to leave their summer digs don't lay up huge food stockpiles. Finally, as the birds travel in flocks with other birds that also feed on insects, caching might be a waste of energy, because a stockpile would be pilfered by other birds in the flock. Overall, food storage is an unlikely key to explain how kinglets survive long winter nights without freezing. The answer to that mystery must thus be elsewhere.

Worker honeybees *flying in and out of the* hive and ranging over some fifty square miles may look as though they are all independent. Yet they are as bound to one another as if they were physically joined. Honeybees (*Apis mellifera*) are social animals and the colony as a whole acts as a single organism and the individuals in it are subservient to the colony's well-being. It is individually in their best genetic interest.

Of the many surprises that have been revealed in these insects over the last century, one that is almost taken for granted is the colony response of regulating its temperature. Even in winter, temperatures in the center of bee clusters remain within a degree or two of 36°C. Whether it's -40°C outside the hive or 40°C, the bees regulate the same hive temperature. Honeybees are the only insects in the Northern Hemisphere that can and do keep themselves active and heated up throughout the northern winter. In winter, they are able to regulate their microclimate protecting themselves and their developing young. Should any one bee leave the communal group in winter, it would, like a cell taken outside an animal's body, die almost instantly by freezing. And if by some miracle it survived the cold, starvation would inevitably kill it. Yet, if a physiologist were to isolate a single honeybee and compare it with any one individual of thousands of other bee species, he or she might not detect anything remarkable. It is only in the context of the colony that much of the marvelous is revealed.

I'll start by considering the highlights of how the bees regulate the temperature of their collective winter cluster, which in late winter and early spring contains eggs and tender larvae. As in overwintering by flying squirrels, kinglets, and most other organisms (including us), choosing the proper shelter or nest site is a primary prerequisite. In honeybees not only must the nest space be adequate for crowds of tens of thousands; it must also be large enough to accommodate nurseries for eggs, larvae, and pupae as well as having huge storage space for a hoard of energy supplies. Choice of proper nest site is important, and when necessary when the colony divides; that choice is not left up to chance.

The colony divides in the spring or early summer when the old queen leaves or is evicted by her daughters if she does not leave voluntarily. The old queen takes with her some 10,000–20,000 daughter helpers. Together the old queen and her many daughters constitute a swarm, which upon leaving the parent colony does not yet have a place to go. The colony at first temporarily clusters on a branch, and from there scout bees then fly forth in search of a new home. Analogous to our own evaluations of a potential house to occupy, bees pace out the dimensions of the place and evaluate other relevant parameters. The scouts then return to the potential site repeatedly, rechecking. Gradually each bee makes a decision, and if she deems the site potentially suitable, then she leaves a scent mark there and then reports back to the swarm.

"Report back" may sound anthropomorphic or overblown. But that's precisely what the bee does. By a series of body movements called the bee dance (a ritualized flight behavior to the potential new home or a good food source), she indicates on the surface of the swarm cluster not only the distance and direction but also an approximation of the suitability of the nest site she has carefully examined. Bees that will soon fly out to examine her indicated site are able to "read" her message because they follow her instructions by dancing with her; by following her body language and accurately deciphering her body movements through their own. Her information is so accurate that others that have never been to the site indicated can fly out and check on it themselves, even if it is miles away.

Any nest search involves numerous scouts, and it usually continues until several scouts have each found potential shelters. Since the whole colony must stay together and thus can go to only *one* nest site, the next obvious problem is that of achieving a consensus on the *best* site after several are indicated. The queen does not make that decision. She has nothing to do with it. She is a follower. Unanimity is reached, instead, first because different scouts check out each other's finds, and secondly because they readily convert if they encounter a better nest site than the one they themselves had found. However, consensus, without the aid of individual intelligence, is reached mainly because the scouts advertising inferior sites stop dancing, while the best sites are advertised more strongly; each potential nest site is advertised honestly according to its relative worth.

A swarm that has left the hive may hang at its temporary home on the branch for hours or days while scouts search for a nest site. During this time the swarm maintains its cluster core temperature near 34° to 36°C, but its cluster-mantle temperature is barely above 15°C (Heinrich 1981). All of the bees on the mantle are too cold to be able to fly, until they warm up by shivering, which costs much energy. The low mantle temperature helps to conserve the swarm's energy supplies, which is critical especially during all of this time they are house-hunting. But how do the mantle bees know when to shiver and get ready to fly? A recent study gives us the answer.

Honeybees' nests were once man's only source of sugar, and humans have maintained an active interest in these insects for this reason alone for thousands of years, as judged by cave paintings, and possibly for millions of years as judged by our sweet tooth and inventiveness to satisfy it. Since early in the last century the bee's inestimable value as pollinators, both in agro- and wild ecosystems, has boosted our perceptions of their lives and their ways even more. With all of the interest and nonstop research on honeybees for over a century, especially that brought to the attention of the whole world by the spectacular discoveries of the dance language by the Austrian Nobel laureate Karl von Frisch and his numerous students and their students and associates and recruits, one would think that the well of discoveries in them would by now be dry.

Not so. Every discovery sets the stage for the next. While von Frisch disclosed to us their stunning dance language as well as their sensory world, neurobiologist Randolf Menzel at the Free University of Berlin deciphered the connections of the bee's senses with their short- and long-term memories. In recent decades we have discovered the mechanisms of how they regulate their individual body temperatures, and how they communally regulate their swarm temperature. Cornell biologist Thomas D. Seeley elucidated how scout bees evaluate the suitability of a potential nest site, which I alluded to previously, and only last year (2001) he with Jürgen Tautz, a colleague from the Theodor-Boveri Institute in Germany, discovered and documented an acoustical "piping" signal bees make by contractions of the flight muscles that is mechanically much like their shivering. Scout bees giving those signals stimulate cool bees on the outermost layers of a swarm cluster to shiver and warm up. Like the bee dance, which can be interpreted as an abstract or greatly abbreviated enactment of a flight toward the *intended* location, the piping signal similarly symbolizes preflight warm-up. As in a warm-up, the sound (vibration) frequency in any one piping sequence starts low (as at low body temperature) and then ends at a high frequency (as at high body temperature). It differs from warm-up vibrations in that bees normally dampen the warm-up or shivering vibrations so that there is little or no sound, as in a smoothly running motor.

Once the cool bees on the swarm mantle are warmed up and capable of flight, another signal, called "buzz running," made also by the scouts, initiates takeoff of the cloud of thousands of bees. It launches them into flight and on their way to their new home. The more synchronized the bees are, the fewer members of the colony will be left behind. The next problem is of bees potentially getting lost from the huge crowd of thousands flying *to* the new nest site, which only a few of the bees in the swarm have seen.

Only some of the bees in the swarm actually know where they are heading. But the swarm crowd is guided by what scientists have dubbed "streakers" (scouts presumably) who zip through the swarm drawing attention to themselves and leading it in the proper direction. The others, and the queen, follow. However, the bees *continue* to fly on only if the queen is with them; they detect her presence by scent.

After the swarm is ensconced in its new nest site, which in the wild is generally a hollow tree, its next main task is foraging for nectar and pollen and building receptacles (out of wax produced from special glands in the abdomen) for the pollen and honey that will be hoarded. Honey is the bee's energy fuel for heat production, and pollen or "bee bread" is the primary protein food fed to the young for growth. Honey is an almost pure sugar concentrate, and it is collected as nectar in the bee worker's stomachs and then regurgitated into the wax storage cells.

A colony of north temperate honeybees with ample fields surrounding them can produce nearly two hundred pounds of honey in a single summer. This is more than enough fuel to keep the hive warm all winter, so that we can in good conscience collect and eat the hive's surplus without destroying the colony. Two or threee hives are sufficient for providing all of my family's sugar needs for a year.

Honey, as such, of course gives off no heat until it is combusted (burned), or metabolized. In both cases the retrieval of the energy from the carbon-carbon bonds of the sugar molecules requires oxygen and yields carbon dioxide and water. Honeybees metabolize honey (i.e., the sugar in it) during flight while collecting pollen and more honey, and exclusively to get heat during shivering. Shivering involves the same muscles used for flight, only the wings don't move because the upstroke and downstroke wing muscles are each in a slow tetanus contraction, one pulling against the other until both are pulled taut.

Shivering, since it uses up valuable honey stores, is minimized by the bees if it can be. Instead, their first response in the winter cluster, which is much like a swarm cluster, is energy conservation. As temperatures outside and then within the hive get lower, the bees begin to draw closer toward one another to form an ever-smaller and -tighter cluster.

The overall effect of cluster temperature regulation can be explained by distinguishing the bees on the outside of the bee cluster—the mantle bees—from those in the center of the cluster—the core bees. The lower the external temperatures, the more the mantle bees try to crawl *into* the cluster. When the cluster has shrunk to near minimum size, then the outermost mantle bees can finally only force their front ends inside. They plug every hole, and heat produced by bees' metabolism within the cluster becomes trapped. Of course, some heat still leaks out by convection and conduction, and for a while the mantle bees are still sufficiently warmed by it. But eventually the mantle may become cold enough for the bees in it to finally have to shiver, to rely on heat produced on their own. In contrast, when external temperatures rise, then there is a lesser temperature gradient from the inside to the outside of the cluster. Less heat then leaves, so the bees inside may start to heat up. After becoming too hot, they crawl out to where it is cooler, onto the mantle, creating cavities and holes through the cluster. As a result of the bees' escaping the heat, the bee cluster then expands, more air channels are created through it, and even more heat leaves.

No central control is required to achieve this communal response that automatically stabilizes the cluster microclimate. No chemical signal from the queen in the center serves as a thermoregulatory directive, since groups of bees with and without their queen react similarly. Neither do bees carry messages back and forth from outside to inside, because when the core bees are experimentally separated from the mantle bees by a thin gauze there is also no temperature change. That is, even when bees of the core are prevented from individually sampling the temperature surrounding the cluster, the core still maintains the same temperature. I also found no change in bee cluster temperature when I played back recordings of buzzing sounds generated by bees either in the core or the mantle at high or low temperature through a small speaker placed in a cluster core or its mantle. Additionally, I exchanged air by pumping it from the core to mantle, and vice versa, and also found no effect. Apparently, the bees in the core and the mantle do not inform each other about local temperature in order to coordinate a colony response. Instead, the high and relatively stable core temperature and the lower and more variable mantle temperatures can both be reasonably well explained by the response of the *individual* bees regulating their own temperature. The summed response, however, serves all in terms of energy economy.

Sometimes, however, the individual's *sacrifice* is best for the colony, and given that bees are communal animals and the individual sterile workers (who are all female) produce offspring only indirectly through their siblings (new queens), that sacrifice is then selected through evolution. The most obvious sacrifice workers make is their attack on nest predators by stinging them. (When a honeybee worker stings just once, its barbed stinger is detached from its body along with the attached poison sac, and that bee soon dies.)

Worker bees may sacrifice their lives in another way. They may die not only by losing body parts, but also by quickly freezing to death. At least so it seemed to me when I watched my hives in the 2000–2001 winter. As is to be expected when examining something of sufficient complexity and interest, I made false starts, but learned along the way. Like the CIA.

During the Vietnam era, GIs discovered mysterious yellow spots on jungle foliage. The CIA was brought in to investigate the so-called mysterious "yellow rain," which was soon suspected to be a new chemical weapon sprayed by the Vietcong. But entomologists later revealed that the smelly yellow mystery droplets came . . . from bees.

Yellow spots are much more visible on white snow than on jungle foliage, and everyone can see them in the winter near their beehives here in northern climates. One also sees bees flying out of the hive in the winter whenever there is a thaw, and the prevailing wisdom is that they leave to take a poop if that is what they do. It made sense, just like yellow rain. But that didn't make it true.

Forty thousand or so physically active honeybees exercising all winter to keep warm while crammed into one small space with lethally low temperatures outside face a hygiene problem, just like a bear does while in its winter den. But solutions are found. Bees won't defecate within the hive any more than a bear will, which is to say never. The difference is that unlike a bear in winter, bees eat a lot, and they eat the same food a bear finds irresistible, honey and pollen.

As the hive in winter is always clean of poop (although sometimes littered with corpses), we may see no problem at all, simply because the bees have solved it so well. But how long can they wait? Till spring? Do they die rather than poop? January 2001 provided a unique opportunity to make observations to shed light on the problem, because we had almost daily or at least weekly snowfall interspersed with sporadic bouts of sunshine with temperatures rising all the way to 2°C.

We had our first letup of the bitter cold at the end of the first week in January. The weatherman talked of "the January thaw." It was warm enough so that the kids could get out and play in the snow without running back inside within two minutes crying from the cold. Kids were not the only warm-blooded creatures who ventured out, at least briefly. Some of my bees did so as well, and bees are decidedly more vulnerable to cold than kids, given their huge size-disadvantage for heat retention.

The bees had by then already been confined in the hive for about two months. Admittedly, they had been fueled with honey, a relatively clean-burning fuel. But the honey also contains small amounts of impurities (such as amino acids) that result in bulk uric acid wastes accumulating in the gut. The thaw came just in time, I thought, for some cleansing flights.

At the first hint of warmer weather, my honeybees flew out quickly and some of them left yellow stains on the snow. They also left corpses; not all bees survived the hazardous outdoor ventures to void their bowels: The snow in front of my three hives under the roof against the side of the chicken pen was pocked with fifty-three seemingly dead bees.

Unlike other insects here in the north, honeybees cannot survive freezing. They must keep body temperature above about 15°C at all times to be able to stay active (or crawl), and they need a muscle temperature of at least 30°C to operate their wing muscles to generate enough power to achieve lift for level flight. Within the hive in winter, many bees tolerate their body temperature dropping to 12° to 15°C. These temperatures are their lower tolerable limit for walking, and it is also the low body temperature set-point that they "defend" when on the swarm or bee cluster mantle, because at still lower body temperatures they become immobile. At body temperatures below about 12°C they start losing physiological control. They cannot shiver to warm themselves back up by their own metabolism, and they are then also unable to crawl back into the social cluster to be warmed by the collective metabolism of the colony. Thus, if a bee should get separated from the winter colony cluster, it would die as soon as the temperature surrounding them (such as on snow) is about 3° to 4°C lower than 0°C since once outside the hive, and at a body temperature of -2°C, internal ice-crystal formation kills them. When honeybees fly out to void themselves at near 0°C, they can therefore not drop their body temperature appreciably before they must return to the warmth of the bee cluster in the hive.

Honeybees are only capable of maintaining a modest difference (of about 15°C) between body and air temperature by shivering and/or flight metabolism; thus if they leave the hive into 0°C air, they are in mortal danger. The bees' corpses littering the snow in front of my hives were of those individuals whose body temperatures had first declined to below 30°C, and then plummeted to lethal temperatures because they lost the ability to control their body temperature. That is, they were those that lost power, fell to the snow, and then cooled to -2°C or less, to freeze solid.

Do the bees experience a reluctance similar to my own to go to the outhouse at my Maine cabin on subzero nights? And when they do fly out do they try to be as brief as possible, since to tarry even a minute means to court death? Do they take off as hot as possible, to thus increase flight speed and delay the inevitable cooling to lethal low temperatures?

Given the bees' small body size, cooling occurs in seconds. Caution in leaving is likely an important variable for natural selection and caution differs among different populations. I found previously that African honeybees (the so-called "killer" bees), for example, are at least as able as our native (European) honeybees at regulating their body temperatures through shivering. Yet, as soon as the African bees reach northern latitudes they suffer huge colony mortality because their workers are insufficiently inhibited from leaving the hive into the cold winter air over the dangerous icy blanket of snow that they have not experienced in their evolutionary history. Unlike the northern honeybees, these bees incautiously rush forth into cold, whereas the northerners proceed much more reluctantly. However, I thought that my bees proceeded not reluctantly enough, even at 1° to 2°C.

Two days later temperatures had dipped to -7°C, and no bees whatsoever were then venturing out spontaneously, providing an ideal opportunity to test their physiological limits. Poking a twig into the hive I generated a crowd of bees rushing to and congregating at the entrance. After a little more poking, I succeeded in provoking a few individuals to fly at and around me. I timed the flight durations of ten of them. Most nose-dived into the snow in just 2 seconds. None lasted over 6 seconds (the mean was 3.6 seconds) before it hit the snow, buzzed briefly, stopped moving, and then froze solid. Between takeoff and landing at flight cessation, the bee's thoracic temperatures declined from 38° to 40°C at takeoff to 29° to 33°C (the mean was 30.9), right after they crashed when I grabbed them and determined their thoracic temperature with my electronic thermometer by the time-tested "grab and stab" technique. (Head temperatures would have been several degrees Celsius lower.) None made it farther than fifteen feet from the hive. After these measurements I agitated thirty-three more of them to leave the hive to see if any could make it back. None did. All fifty-five bees that I examined had the foul-smelling yellow paste in their rectums. My crude experiment showed that if they do leave at -7°C, which they were only willing to do in defense of the hive, they risk certain death. At what temperature would they risk flying out on their own?

On January 20 we had sunshine in the afternoon, although air temperatures remained low, near -9°C. But at about 2:30 P.M. the sun hit the hives broadside and bees started coming out spontaneously. In the half hour that I watched as an innocent bystander, 125 flew out. Every single one of these bees took its (involuntary) kamikaze dive into the snow after a few seconds of very rapid cooling in flight. Not one of these 125 eager leavers made it back into the hive. All solidified from internal icing. Most (112) of these bees had not voided their rectums, as I could easily ascertain visually by pulling open their abdomens. I was puzzled now, wondering: Why didn't they defecate when they had the chance? They could not have flown out to commit suicide! Why had they risked death? I was determined to get to the bottom of this mystery. The next day, when it was warmer, I got another clue.

There was again a short period of sunshine in the afternoon. This time over three hundred bees had come out spontaneously—I found them dead and strewn all over the snow in front of the hives. All had died there, as before, since the snow was still near -6°C and so they all froze solid. Also as before, there was only modest fecal speckling on the snow; there was no soiling like one might expect if hundreds of bees had rushed out to relieve themselves. Were those bees at the entrance, the ones that rushed out, the subpopulation that most urgently had to relieve themselves but that died from cold before they had a chance? With that question in mind, I lifted a hive cover (the point in the hive farthest removed from the entrance) and retrieved fifty bees from the bee cluster just underneath it. All fifty of these had feces in their rectums, and I could detect no difference in the amounts they had from those in bees that had risked flying out into the cold. Neither did it seem that the abdominal temperatures of those outside had been too cold to perform their function;

the abdominal temperature of those hitting the snow was still relatively high (11° to 19°C with a mean of 12° to 15.5°C).

Apparently the bees had not come out *just to defecate*. I also doubted that they failed to get back into the hive *only* because of the cold since many of the bees hit the snow in fully powered dives. Might the bees (who had of course never seen snow before) have been disoriented by not seeing dark ground under them, and so they had accidentally dived into the powdery white and featureless snow? To find out I now spread wood shavings all around the hives, and I then provoked more bees to come out. Air temperatures were -8.3°C. Voila! Although many bees crashed as before, sixteen out of forty now made it safely back inside, whereas none had done so before even at considerably higher temperatures. Therefore, at least some of the earlier deaths were probably due to disorientation over the light featureless snow that slowed their ability to get back.

On January 22 of that year we had a thaw and temperatures finally reached a few degrees Celsius *above* freezing. The shavings were still spread around. This time there were fewer than a dozen new dead bees near the hive itself, but there were at least sixty strewn all over the snow at 100 to 200 feet. Several bees had even flown up to 100 yards beyond the hive before landing in the snow. Why did they fly *so far?* I gathered up as many as I could find, and again examined their bowels. This time twenty-five of them had voided but the rest still retained their fecal load. Unloading is thus something that they do, but it seemed more and more unlikely that the *primary* reason the bees risk their dangerous exiting is for gut evacuation.

By Friday the twenty-sixth, it had again warmed to just barely above freezing, and new snow had covered the many dead bees so the new corpses could be counted. I counted 225 bees that had recently crashed into the snow in the hive area. I then watched my hives for forty minutes in early afternoon. Many bees were still coming out spontaneously. I kept my eye on individual bees, noting their typical back-and-forth orientation flights, circling, and then their departure into the distance till they were out of sight or till they crashed! Of 171 bees that I followed visually, 96 crashed into the snow, while 52 went out of my vision into the distance. I saw only three defecations, finally convincing me that although the bees proximally "cleanse" on some of their so-called "cleansing flights," the *ultimate* reason for those flights had to be something different. Could the flights be scout bees searching for flowers to get food, or water to drink?

On New Year's Day I had picked willow twigs and brought them into the house. In two weeks male catkins were shedding pollen. Quaking aspen buds brought in at the end of January opened in only four days. It was not yet spring, but the plants were ready. The bee cannot know if or when the red maple, willow, and poplar trees in the surrounding bogs and forest will burst forth with their very brief one-time offerings. The colony can't know by divine inspiration, and it cannot afford to miss the early spring harvest. But it can afford to lose some workers to buy information.

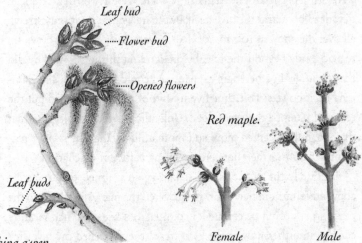

Leaf bud

Flower bud

Opened flowers

Red maple.

Leaf buds

Quaking aspen.
(Populus tremuloides)

Female

Male

There is a huge premium for swarms to leave the colony *early* in spring, to allow for sufficient time to find a new nest site, build honeycomb, rear young, and build up the large honey stores required to get them through the winter. The only way for the bees to get information on when the first blooms are available, so that swarms can be launched sufficiently early to accomplish all this, is to venture out and search. A few hundred, or a few thousand, worker casualties may be a small price to pay for being at the first bloom (or being first at the bloom). After all, the colony is a superorganism whose success is measured by the reproductive output of one *queen,* and the queen's output depends on honey and pollen input.

The fifteenth of March was finally a "warm" (8° to 10°C) sunny day. On this day I saw a first no-doubt-honest-to-goodness *cleansing* flight: On this day the air was loudly abuzz with thousands of bees at any one time, and on this day it fairly rained yellow droplets continuously. I cleared five 1-square-foot sections of snow and found 80, 95, 94, 102, and 160 fresh droplets collected on them, respectively, in just thirty minutes. There were almost no dead bees on the snow. Bees repeatedly landed on the snow, but they all got up to fly again. Most bees vanished from sight before again returning. They were foraging. They were seeking flowers, and the wind-pollinated trees from which they gather a major pollen crop bloom early. The poplars and red maples open their blossoms in late March. No other pollen is then available, but I counted up to 154 bees laden with poplar pollen returning to one hive per minute on the day after the poplars first bloomed, and when winter was not yet over. (There are then still frequent nightly frosts, and in 2002, the next year, a storm dumped six inches of snow on April 22, three weeks *after* the poplars had finished blooming.) The bees' sacrifices paid off. All the colonies survived the winter, issued swarms, and made near-record amounts of honey that spring and summer.

By the end *of January, those of us who live in* the Northeast have been looking out on starkly bare trees for three months. "Only four more months," we think, before the buds break and trees are again resplendent in green. The wait is all the more difficult when you realize that the buds are there all winter, biding their time. Indeed, they were fully formed on the trees the previous summer, long before the shows of brilliant fall foliage.

A bud can consist of bare clusters of miniature leaves harboring a new shoot (as in hickory and butternut); an embryonic flower or inflorescence (as in alder, hazelnut, and birch); or both incipient leaves and flowers encased together under protective scales (apple, cherry, shadbush). The large buds of mountain ash and poplars have a sticky, resinous covering that helps protect them from

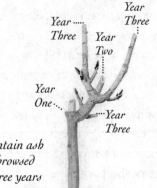

Red maple shoot with deer browse and an empty sawfly cocoon.

hungry animals. All leaf and flower buds are packed with nutrition and are prized winter food for many northern herbivores. Ruffed grouse live all winter on aspen and birch buds, and I've watched purple finches devouring sugar maple flower buds. I've seen red squirrels decapitate almost all the young balsam fir trees in some patches to eat the large terminal shoot buds. In the tops of mature fir trees they snip off hundreds of terminal twigs to eat the twenty or more flower cone buds on each one like corn from a cob, as shown in the figure on page 44. Then they discard the twig, which drops onto the snow below. Moose break off branches of poplar saplings and of red and striped maple to feed on the termi-

Year Three

Year Three

Year Two

Year One

Year Three

A twig of mountain ash that has been browsed off by hares three years in succession.

nal buds and twigs. In some patches of my woods in Maine, I can hardly find a single sapling that is untouched by moose, deer, or snowshoe hares.

Buds vary greatly in size, and it is primarily the trees of northern regions that have *large* buds. Southern transplants to the north, such as black locust (*Robinia pseudoacacia*) and honey locust (*Gleditsia trichanthos*), have only miniscule buds.

For large-budded northern trees, the prepackaging of leaves and flowers into buds must have some advantages that outweigh the considerable cost of maintaining buds for so long before they are activated. I suspect the main benefit of having the leaves and flowers preformed all winter is for a quick start-up in the spring—the buds are ready to break out quickly on cue, thereby allowing the tree to make the most of a short growing season. In New England, trees have only three short months in which to produce leaves—their photosynthetic machinery—and then use them to make an energy profit. When the long-awaited spring finally arrives in the north, it does so suddenly. Over a week or two, the ice and snow melt, and the bare ground begins to absorb heat. In only three or four days in mid-May, a bare beech forest is crowned with a canopy of pea-green leaves soaking up sunlight.

Pin cherry, with flowers and leaves from same buds.

Butternut, with separate leaf and flower buds.

····*Leaf*

····*Flower*

There is, however, a major caveat in the trees' race to grab light: the new leaves are vulnerable to frost damage. Buds, as long as they are dormant, can, like hibernating insects, survive winter's lowest subzero temperatures. Once they awaken and begin to draw water into their tissues, however, they are at risk. They in turn can put the tree at risk, because leaves can collect wet snow that can break the tree branches. It's a dilemma. The buds need to open as early as possible, but not too early. Trees, being long-lived, can perhaps afford to lose flowers to frost in any one year, because energy saved one year by not fruiting can instead be invested in growth and can lead to the production of even more fruits the next year. Losing leaves to frost, however, is more serious; if the tree misses out on the sunlight sweepstakes, or loses limbs to snow-loading, growth and reproduction will suffer. Releafing is sometimes possible but is energetically costly. However, trees are seldom fooled by a false start, such as a midwinter thaw. How do they know when to start their metabolic engines and break bud?

Buds follow local schedules that are dictated by an interplay of cues involving day length, seasonal duration of cold exposure, and warmth. Warmth alone is not enough. For example, sugar maples from the north, if transplanted to Georgia, won't break bud there because they need a long period of cold beforehand, a kind of reminder that winter has occurred. The strategy is a bit like that of northern cecropia moth pupae, which do not stir unless first chilled for a sufficiently long time.

Where I spend most of my time, in western Maine and in central Vermont, new leaves of all the deciduous tree species usually emerge relatively synchronously in mid-May, over the short span of about two weeks. First to appear are quaking aspen and birch leaves; last are oaks and ash. Beech, maples, and the others are in between. *Flower* buds of the native forest trees, however, open in a progression and over a six-month span, starting in March or April with poplars, alders, red maple, and beaked hazel; moving on to basswood in June and American chestnut in late July; and ending in October with witch hazel. (Significantly, the latest-blooming species do not have their flowers prepacked in buds in winter.) In trees such as apple, cherry,

*Willow showing swelling of
flower buds, but not of leaf
buds, in response to warmth.*

·····*Unopened
flower buds*

·····*Leaf buds*

and shadbush, with flowers and leaves packed in the same bud, the flowers generally bloom in one quick burst; the leaves follow almost immediately.

Having separate buds for leaves and flowers—such as in willow, poplar, and alder—allows a tree to open its flower buds a month before the leaf buds or up to five months after them, such as in witch hazel. Wind-pollinated trees may produce flowers a month or more before leaves, which tend to block wind flow. In contrast, a bee-pollinated basswood may flower a month or more after the leaf buds have opened, when bee populations peak in late summer. Witch hazel takes advantage of the pollination services of winter moths of the genus *Eupsilia*, which are active in fall and winter (see Chapter 14).

Bud opening is a wonder but can easily be taken for granted. I like to be reminded of the spring miracle, especially in the depth of winter, when the vibrantly alive trees look so dead. Every year in January, February, and March I go into the woods, pick some twigs of trees and shrubs, then bring them home and stick them in a jar of water. Indoors, some buds can be coaxed (or "forced," according to botani-

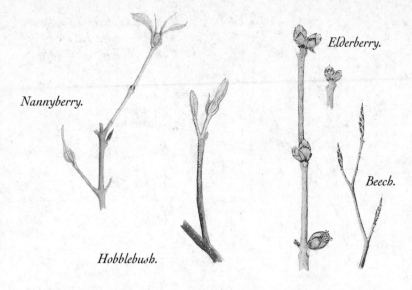

Nannyberry.

Elderberry.

Beech.

Hobblebush.

cal usage) to open at least three months ahead of their normal out-door schedule. Twigs of trembling aspen, willow, beaked hazel, speckled alder, and red maple, picked as early as January and brought inside, will flower and then shed their pollen. (The flowers of these trees and shrubs are also the first to open in the woods, in early March or April, when there is still snow on the ground.) In contrast, most leaf buds, as well as the flower buds of late-blooming trees such as basswood, don't respond to the warmth of my office until March.

When snowstorms rage outside and temperature may dip below 0°F (-16°C), the jar of twigs reminds me of the winter trees' vibrant life. The buds on the trees outside are like runners who have prepared for a big race for more than six months and who now wait for the signal to start.

In late April or early May, a warm temperature pulse will be the "go" signal, starting the twigs and leaves on their race to the sun-light and summer. In 2002, that pulse, two days of 90°F (34°C) on April 19 and 20, was unusually early. It forced the poplars, cherries, serviceberries, and sugar maples to start unfurling their leaves, to be followed two days later by the flowering of the serviceberries and sugar maples. Two snowstoms and frosts then followed within ten days. A week later sticky snowflakes the size of miniature snowballs were plummeting down and sticking onto the leaves and flowers. It looked like the trees had made a false start—as though they had jumped the gun. If so, they would be penalized themselves. This time they seemed okay, but I think they were cutting it close. The costs involved only a few broken limbs, a few lost leaves. As in the bees' early exits from their hive to forage, the gamble paid off.

Kinglets are small, beautiful, and pure in
their simplicity. They remind me of snow crystals.
Each snow crystal is a six-cornered starlet formed
according to the unvarying laws of physics and
chemistry. Each one is perfect. Yet the diversity in
their shapes is astounding in part because any tiny
random event in their formation shapes all future
events in their growth.

Similarly, in the evolution of organisms and
ecosystems there are innumerable random events
of history that shape outcome. Outcome is not
preordained. There are no correcting factors or laws
from above that *specify* the form of the ultimate
outcome, even as the shape of the kinglets' adaptations necessarily conforms to energy economy of
life just as the snow crystals' shape conforms to the
blind energy economy of physics. Ultimately, the
kinglet is to a snapping turtle, or a crossbill, or an

arctic ground squirrel, or to us, as one snow crystal is to another. But to appreciate that fact requires seeing them up close.

I made my first attempt to catch a kinglet in my hand when I was about nine years old, and I remember the occasion clearly. It was in the late fall, and I was alone, walking along a woods road back from the village school, when I encountered a small group of them foraging close to the ground among some young spruce trees. I got almost close enough to touch them. They seemed oblivious to me, and I reached out to try to capture one. I did not succeed, but they had captured my imagination, as they have many others'.

In the introduction to the golden-crowned kinglet chapter in his now truly classic multivolume *Life Histories,* a standard reference text on North American birds, Arthur Cleveland Bent wrote: "Many years ago, a boy found on the doorstep the body of a tiny feathered gem. Perhaps the cat had left it there, but, as it was a bitter, cold morning in midwinter, it is more likely that it had perished with the cold and hunger. He picked it up and was entranced with the delicate beauty of its soft olive colors and with its crown of brilliant orange and gold, which glowed like a ball of fire. In his eagerness to preserve it, he attempted to make his first birdskin. It made a sorry-looking specimen, but it was the beginning of a life-long interest in birds, which lasted for half a century."

Kinglets are named *Regulus* ("little king") for their bright lemon yellow, orange, and red crowns. In the golden-crowned kinglet (*Regulus satrapa*), the females' crown is yellow and the males have an orange-red crest of feathers within the yellow crown that is usually hidden out of sight, but it is raised like a flickering flame to show excitement.

Taxonomically, kinglets have long been a puzzle. Bent placed them among the thrushes and their allies. They have also been assigned to the titmice family, Paridae. They were thought closely related to the Old World warblers, subfamily Sylviinae (as opposed to the New World warblers, Parulidae). However, DNA studies (Sibley and Ahlquist 1985) suggest that they are of another descent (Ingold and Galati 1997); they are now known to be unrelated to either thrushes, tits, or warblers (Sheldon and Gill 1996). They are different, and unique.

In Europe there are two species of kinglets, the goldcrest *Regulus regulus* and the firecrest *Regulus ignicapillus.* Both species are relatively sedentary, but not entirely so as they have a tendency to migrate north and south within Europe. In North America there are also two species of kinglets. The ruby-crowned kinglet (*Regulus calendula*) breeds in Alaska and throughout Canada and migrates in the winter to Mexico and the southern United States. In contrast, the golden-crowned *Regulus satrapa* occurs throughout the United States and southern Canada in winter and summer, although part of the population also migrates into northern Canada to breed there.

Despite the depictions in most bird books, recent studies of the kinglet's proteins (Ingold, Weight, and Guttman 1988) indicate that there are substantial genetic differences between the two American species. These differences are large enough to warrant putting them into different genera. On the other hand, our North American golden-crowned kinglets are nearly indistinguishable from the European and Asiatic goldcrest. Even their songs are nearly identical (Desfayes 1965). They could easily be lumped into the *same* species, although for perhaps no other than practical purposes they currently aren't. For practical purposes I also restrict my discussion to the North American golden-crowned kinglet.

The golden-crowned kinglet's ability to hold on in the north, where insect food is plentiful in summer and scarce in winter, is in no small measure dependent on its nesting behavior. Kinglets produce many young that compensate not so much for a high nestling mortality as in most other northern songbirds, but for a high winter mortality. Their nesting behavior is noteworthy in several ways. First,

Golden-crowned kinglet beginning to build a nest.

even in Maine and in Nova Scotia, where the deciduous trees don't leaf out until mid-May, the kinglets already start nest-building in mid-April. Snowstorms are still common at that time and if kinglets built their nests like many other birds, then these nests would often be buried under snow. However, unlike crossbills, kinglets build their nests suspended *under* spruce branches where they are covered from above by a thick latticework of twigs and needles. Here the nest is virtually invisible from above and therefore well protected from most predators, and if the branches over a nest get covered with a thick layer of snow then all the better, because the nest is then inside a snug insulating snow cave.

Cold is still a potential problem early in the year, but a kinglet's nest is built for warmth. Few people have observed the nest-building process, and even fewer have described it. I quote here a detailed set of observations of Miss Cordelia J. Stanwood, of Ellsworth, Maine, whose notes were published by Bent (1964, p. 386):

> The kinglets selected for the roof of their cradle a heavy spruce limb with a dense tip; and the female, hopping down through the branch from twig to twig, attached her pensile nest to the sprays.
>
> The bird wove her spherical structure about herself much as the caterpillar of the luna or cecropia moth weaves its cocoon about itself, except that the kinglet had to gather her materials. The bird stood on a twig on one side of the space she had chosen for her nest and measured off her length, as far as the situation of the twigs

would permit, by attaching bits of spider's silk and moss to the twigs. Thus she laid off the points for the approximate circle for the top of the nest. Then she spanned the space through the center of the circle, roughly speaking from north to south, with spider's silk and moss, forming a sort of cable, which later assumed the appearance of a hammock. After a time, when the bird came with moss or silk, she would fly down upon the hammock as if to test its strength and lengthen it. At all times, however, she worked all over the nest from left to right, moving her beak back and forth as she secured the silk and moss and stretched the web from one point of attachment to another. As soon as the hammock would support the bird, she stood in the center and walked around from left to right. When the hammock was wide enough to admit of her sitting down, she modeled the center of the suspended band by burrowing against it with her breast, and making a kicking motion with her feet. Gradually she embodied some of the twigs in the structure, as if for ribs, and occasionally she snipped off a spruce twig to use in shaping the globular nest. At last the bottom, or basketlike part, arose to meet the top of the nest and the industrious goldcrest was hidden from sight as she labored.

The creation was really a silken cocoon, in the walls of which was suspended enough moss, hair, and feathers to render it a nonconductor of heat, cold, and moisture. This primitive incubator was made of the same fine, dark yellow-green moss, *Hypnum uncinatum*, that seems characteristic of the habitations of the golden-crowned kinglet in this locality, *Usnea longissima*, a long, fringelike lichen, and animal silk. More of the gray-green *Usnea* lichen was used in the hammocklike band around the middle of the nest than in other parts of the well-made structure. The lining consisted of rabbit hair, I think, and partridge feathers. The wall of the abode was all of an inch and a half thick, and the window in the roof measured an inch and a half in diameter.

Apparently the female alone builds the nest. The male accompanies her and sings as she gathers nesting material and builds. Bent (1964) provides descriptions of other golden-crowned kinglets' nests from the Northeast, and he documents the birds' habit of lining the nests with feathers. Ruffed grouse ("partridge") feathers appear to be used in almost all nests. The breast and body feathers of grouse are of course rather huge for the small kinglet's nest, but the birds insert these feathers into their nest lining so that the quills point down into the bottom of the nest and the natural curve of the feather reaches over the top, forming a soft flexible curtainlike cover.

A kinglet incubating all of its eggs, with the help of its hot feet.

The majority of songbirds breeding in New England have four to five eggs per clutch. Kinglets have a whopping eight to eleven in a single clutch. They lay so many eggs and their nests are so small that, unlike other birds, the eggs are in two layers, usually five on the bottom and four on top. The bee-sized kinglet hatchlings are pink, blind, and naked. Stanwood (in Bent 1964, p. 388), in describing the young, says:

> At the approach of the parent birds, they raise their little, palpitating bodies and open wide their tiny, orange-red mouths for food. These mouths are about the color of the meat of a peach around the stone. The veins showing through the thin skin give the bodies much the same tone. At first the young are fed by regurgitating partly digested food; later moths, caterpillars, and other insects furnish their diet. They are very fond of spruce bud moths and caterpillars. A beautiful triple spruce was attacked by these pests and

almost denuded of its foliage. I noticed the kinglets frequenting this tree a great deal. In a season or two, the foliage was as luxuriant as it had been in the past. Such are the good offices performed by the golden-crowned kinglets and their young. The feet of the young are large and strong for the size of their bodies. If a person attempts to lift one from the nest, the little fellow will tear the lining out before he will release his hold. Just before the feathers appear the young begin to preen, and after that spend much of the remainder of their time in the nest smoothing and oiling their plumage. The parent birds remove all waste, depositing it far away from the little home, which is kept clean and sweet.

She continues:

I have seen kinglets feeding young in the nest as late as the last of June, but by the eighteenth or twentieth day of June, goldcrest families are usually foraging in the trees. As late as the middle of September in 1912, I saw mature kinglets industriously feeding a large family of young birds in a seedling grove.

At the time that Cordelia Stanwood was making her observations, nobody had yet discovered another amazing aspect of the kinglet's already amazing nesting behavior. This only came to light from the loving studies, or studies for the love of the kinglet, by another pair of amateur ornithologists, Robert and Carlyn Galati who also fell for the golden-crowned kinglet. Their observations in northern Minnesota showed that kinglets not only successfully raise a family of nine to ten young in one nest, they are *simultaneously* busy with a second nest with as many young at the same time (Thaler 1990; Galati 1991). The phenomenon is called double-clutching.

As already mentioned, apparently the female alone builds the nest. And she alone incubates (Thaler 1990; Galati 1991). The male is the food provider. After the eggs hatch, the female has to stay on the nest to warm the naked young. The male feeds the whole family. However, soon after the young no longer need to be brooded (which is in part a function of how well-insulated a nest she had built), she deserts her young, starts to build a second nest nearby, and is then soon incubating her second set of eight to ten eggs. Her mate tends the babies of the first brood. It's good bird parenting: despite all the hazards, nesting success is, at over 80 percent, exceptionally high for any bird (Ingold and Galati 1997).

The kinglets' necessarily high death rate, given their high birth rate, results from living close to the energy edge in wintertime and from being weak fliers due to the heavy coat of insulating feathers they wear. Those kinglets that leave on migration suffer enormous mortality (Kania 1983; Hogstad 1984). But there are presumably similarly high losses by *not* migrating, or else migration would soon cease. Those that stay in winter never once stop for even two seconds in their search for food. From early dawn until dark they hop nonstop in their frenetic hunt for insects. Although they can survive nights of -40°C, severe weather and insufficient food to fuel their metabolism may produce 100 percent mortality in severe storms and icing (Lepthien and Bock 1976; Larrison and Sonnenberg 1968; Graber and Graber 1979; Sabo 1980).

The golden-crowned Kinglet is one of the three birds featured on the cover of the 1992 book *Birds in Jeopardy* (by Ehrlich et al.), which lists and describes the imperiled and extinct birds of the United States and Canada. However, this kinglet is not so much a rare bird as one that doesn't attract much attention. Kinglets are difficult to see, but even then the kinglet had indeed suffered a severe decline in the early 1980s in some areas. By the end of the decade it recovered. Since kinglets have excellent nesting success and are too small to be preferred prey by most predators, the sharp dip in their population was likely due to a severe weather disturbance. Recovery can be rapid in a species with high reproductive rate, if environmental conditions improve. Kinglets in captivity have a maximum life span near ten years (Thaler 1990), but any adversity can affect them in the wild, where 87 percent of the population is on average normally weeded out every year. Kinglets are as close to an annual bird (in analogy with annual plants that regenerate each year only by seeds) as any birds gets.

For such small insect-eating birds with weak bills unsuited for prying under bark or into wood, winter is a severe enough problem that the alternative of flying thousands of miles over open ocean can be preferable. With millions succumbing annually on these dangerous journeys, there is strong selective pressure on many birds to produce the navigational sophistication and the physical and mental attributes that create and maintain the migratory capacity and behavior. At least some migrants have evolved a new sense that we lack, the ability to detect the lines of magnetic orientation of the earth. Many have evolved to be able to read the star patterns and to navigate by the constant beacon of the North Star at night. In the day they may use the sun as a compass instead. They track its movement across the sky and calculate the movement with the use of an internal time sense that is accurate to within about fifteen minutes; hence they can ascertain direction to within about four angular degrees. Physiologically, birds have evolved cycles of rapid fattening in preparation for days of unremitting flight. And most of these capacities evolved because of winter, without which none of the kinglet's almost unfathomable capabilities would have come into existence. We don't specifically know how golden-crowned kinglets migrate, but since they show nocturnal migratory restlessness (Thaler 1990) they therefore probably fly and navigate at night in order to be able to refuel by day.

Kinglets' evolutionary history, and hence their biology, is linked to the Ice Ages. As recently as ten thousand years ago when the winter snows melted, there was a large, relatively uninhabited portion of the globe where insects became available in almost unlimited quantities for fifteen to twenty-four hours per day for all who would come to harvest the bounty. In the fall, when the insects became unavailable and daylight vanished, the birds retreated south. Gradually as the glaciers melted, the birds' annual journeys to and from the rich northern feeding grounds where the days were long, became even longer. Always it was those that either could stand the cold longest or that could fly the farthest that would collect the largest bounty in the spring. They, on average, left the most offspring. Parts of the population become reproductively isolated through the glaciers of this last and previous Ice Ages. Variations, which we often arbitrarily call species, were then created. One successful group was the kinglets, which now occupy the taiga forests of the north.

Few people ever get to see a golden-crowned kinglet, even if they are looking for them (and granted, most of the population is not looking for kinglets). Golden-crowns are very difficult to see, even without the dense cover of coniferous forest in which they live. Given that sightings are difficult, the best way of determining their presence is by listening for their calls. I'm happy to report that at least at the present time, this "little king" of the forest is doing well in the dense coniferous regions of northern New England. It is rare indeed that I do not hear their pleasant *Tsees* in the winter woods near my cabin. The wonder is how they survive a winter night.

Whenever I return to my heated cabin on a winter day, I can be secure in the knowledge that I won't freeze to death. Our species has a magic key for winter survival. That key, as Jack London's story told, is fire. Other human species, like the Neanderthals, also possessed that key for probably hundreds of thousands of years. Without it, humans would not have colonized the Northern Hemisphere of the globe across Europe, Asia, and North America, all the way up to the edge of the glaciers. Fire has not only helped keep us warm and alive in the night; it also allowed us to be better predators because by cooking our meat we use it as food more efficiently. And when we were still prey, fire was also a weapon for defense.

The kinglet has occupied the same circumpolar realm that we have, and it has likely done so for incomparably longer periods of time. It has the same requirement for heat, but raised to a much sharper edge. Given its minute, twopenny weight (5 to 6 grams), how such an individual could survive the energy crunch on a cold, sixteen-hour-long winter night is an unimaginable marvel from our human perspective—it defies physics and physiology. We don't know for sure how they do it, so we search for explanations offered by animals, and especially other birds. Their example suggests that the kinglet's winter survival will not likely involve a new biological phenomenon or new laws of physics and physiology; I suspect, rather, it will depend on a species-specific balance created by precisely juggling a set of conflicting benefits and their costs and doing everything just right with little margin for error. There is no magic. It's a matter of details—of getting everything just right.

I was attracted to them in part because I understood so little, and to some extent I don't mind keeping it that way, to preserve the mystery. My pursuit of hard fact is not for the sake of facts. It's to "capture" the story behind them.

Thaler (1990) presumed that the European goldcrest survives in winter by huddling at night to save energy, fueling itself on an energy base of springtails (Collembola), and by its ability to seek microclimates such as under cushions of powdery snow and other places. However, there is no energy balance sheet to determine the limits of these strategies, and although other birds tell us where to look, what they do does not necessarily apply to the golden-crowned kinglet which, given its North American haunts, faces even lower temperatures than its European relative. Our studies of foraging behavior of kinglets in the Maine woods (discussed in Chapter 9) showed, surprisingly, that they either did not have or did not use springtails as their main energy source. Instead, stomach contents showed that they subsisted on tiny frozen caterpillars of the moth family Geometridae. Caterpillars are not of high caloric content, as seeds are, so these little insectivorous birds would like many others drop their body temperature at night to become torpid. Nevertheless, it was presumed that reduction of body temperature at night is *not* necessary for winter survival in goldcrests (Reinetsen and Thaler 1988). This assessment was based on an overnight weight loss of 1.3 to 1.5 grams that was assumed to be fat. However, the birds could have a much lesser fat cushion than that, since much of the overnight weight loss could have been gut contents. Charles R. Blem and J. F. Pagels examined fat, specifically, and showed that in midwinter the extractable fat reserves accumulated by the North American golden-crowned kinglet over a day were about five times less. These authors calculated that this amount of fat (0.3 gram) would contain *insufficient* calories for a kinglet to maintain a high body temperature all night.

I think we vastly underestimate these, and many other birds, when we expect them to follow simple rules. Kinglets probably won't go torpid unless they have to. As can be inferred from innumerable studies of other animals, what happens in captivity (when the animals are well fed and not stressed) may be a pale reflection of what they confront, and perform, under field conditions.

The kinglets' main adaptations for keeping warm (and conserving energy) include those also found in most other birds. They fluff out their feathers to trap air, creating an ever-greater insulating air space around themselves. The main avenue of heat loss is then through the uninsulated bill, eyes, and the feet. During sleep, however, the first two avenues are greatly reduced as the birds tuck their heads deep into their back feathers. Reductions of body heat loss through the feet is accomplished by countercurrent heat exchange and/or reduction in blood flow, to keep leg and foot temperatures as low as possible, probably just above the freezing point of water, near

0°C. Conversely, the kinglet's legs and feet can also be used to shunt heat *from* the body. For example, Thaler (1990) observed that kinglets normally have light brown legs, but when incubating their eggs, females' legs blush to pink and red as blood is flushed through them and leg temperatures reach 39°C. (The kinglets' broodpatch on the chest and belly is only sufficient to simultaneously incubate two to three of the up to eleven eggs of one clutch, and the heated legs are needed to constantly shuffle the eggs, and incubate them.)

In Jack London's story "To Build a Fire," the old-timer on Sulphur Creek had told the cheechako that "no man must travel alone in the Klondike after fifty below," or as Alaskans quip when it is very cold "it's a two- [or three?] dog night." Similarly because of the danger posed by cold, golden-crowned kinglets in the Maine winter woods travel in groups of two to three or more and like goldcrests, they huddle at night. Huddling saves energy. How much heat loss an Alaskan reduces by huddling with a husky has not, to my knowledge, been accurately measured, but one goldcrest huddling with another reduces its heat loss by about 23 percent, while in trios heat loss is lowered by 37 percent, similarly to bushtits (*Psaltriparus minimus*), which also save the same amount of energy overnight by huddling in pairs or trios. Whatever it is that kinglets are known to do, it still doesn't quite add up because the large energy savings achieved by huddling are insufficient to offset differences between energy reserves and energy demands. Even with huddling, hypothermia (reductions in body temperature) in kinglets is likely inevitable.

Deep torpor at night would confer large energy savings. But body temperature is unlikely to be allowed to go much lower than about 10°C, because the birds can't risk losing the ability to shiver to keep from freezing solid if temperatures at night dip to -30° to -40°C. Survival, even with a body temperature of 5° to 10°C, would likely be impossible without the one thing I suspect matters most and that we know the least about: shelter. Migrating goldcrests stopping on a bare rocky island off Scotland, where there was no snow and little vegetation to hide in, were discovered overnighting in the open, often in groups, but many of them died overnight (Brockie 1984).

Pagels and Blem have reported seeing a golden-crowned kinglet entering a squirrel's nest. If golden-crowned kinglets regularly overnight in squirrels' nests, then that should indeed go a long way toward solving their problem. In magnitude, it would be the equivalent of them inventing fire, because it would conserve body heat by enormously reducing convective heat loss. However, squirrels are notorious predators of (young) birds who can't fly, and red squirrels even prey on young snowshoe hares. A tiny kinglet would be a tasty snack for a squirrel. How would a kinglet know if a squirrel nest is uninhabited and worth the risk of entering?

In the winter of 2000–2001, I hunted for and examined dozens of both red and flying squirrel nests in the Maine woods. I became skeptical that these very snug nests (see Chapter 5) could be or were used by kinglets that were active in the same coniferous woods. First, although each nest had two entrances, these entrances were difficult to find. To introduce my hand into a squirrel's nest I had to force it through thick dense nest material; it felt like forcing one's hand through an elastic glove opening that normally stays shut. Could a bird squeeze through? I found no bird feces inside any nest. In all the snow caves where ruffed grouse had overnighted, I found dozens of fecal pellets, and I presume that kinglets entering squirrel nests with a full stomach at night would also have had to void their bowels in the night. I therefore asked Blem for more details. He e-mailed me: "My single observation of kinglets and squirrel nests was very brief, but I definitely thought the bird disappeared in the nest. I could not say that it went into a main entrance. It seemed that it just went into the loose leaves on the outside of the structure."

Since reading about Pagels and Blem's interesting and provocative observation I have been stimulated to stay alert and try to find out where kinglets spend their winter nights, because all indications are that the overnighting roosts are crucial to their survival. I've followed them at dusk, again and again, but always lost track of them as they continued to forage and eventually faded and vanished into the darkening foliage of conifers, usually with no squirrel nest in sight.

In early January 1995, I thought I was finally getting close to tracking them to a sleeping place. I had noticed a group of three of them in the spruces near my cabin, and on January 5 I saw them again and followed them for eighty minutes. Finally, as it was getting dark, I heard them make some persistent soft *tsees,* then many louder ones, and then, at 4:20 P.M., the birds suddenly became quiet and vanished from sight. But it was just getting too dark to see. The next evening I waited near the same area till 4:30 P.M., and saw nothing. A student, Jeremy Cohen, took over near the same site the next three evenings, and he managed to follow a (or the) group of three kinglets on one evening, again until 4:30 P.M. when it was almost dark. None had entered the three local red squirrel nests that I had located near there previously. At dawn the following day, I arrived with Cohen and other trackers at the same site where they had been last seen on the evening before, and we did indeed see three birds just as it was getting light. But it was not near a squirrel nest. I then became ever more doubtful that the kinglets' key to winter survival in the Maine woods could be traced to squirrel nests.

I also doubted that kinglets would routinely burrow into the snow on the ground and dig tunnels like grouse do. That is because I've frequently encountered rainstorms followed by icing that produced thick crust on the snow. Only a large strong bird can escape the icy prison, or last it out till the ice melts. It seemed possible, however, that kinglets could burrow into the *undersides* of snow cushions on branches, and in that case they could again escape from below, because the ice crust forms only on top.

In the winter of 2000–2001, I and my Winter Ecology students again made it one of our projects to try to pursue kinglets to their sleeping quarters. Again we were unsuccessful in tracing any birds into a squirrel nest. Nevertheless, we probably got closer than ever before to what they actually do. It was by accident.

*Two kinglets fluffed out and huddling
in a snow cave on a branch.*

On December 19, one of the students, Willard Morgan, went out before sunrise to enter a hiding shelter we had built out of brush to watch ravens (where they could not watch him) arriving at a carcass. On his way to that shelter, while walking in the semidarkness, he flushed two kinglets almost off the ground at his feet from under a brush pile he passed. The brush was covered with cushions of freshly fallen snow. The previous night had been windy and blustery, with sleet and snow. Being close to the ground and tucked under the snow cushions in the brush pile, the kinglets would have escaped the weather. Morgan returned to the same site to watch in the next two evenings, but the kinglets did not return. This observation strengthens my suspicion that kinglets, in order to forage to the last minute of the day, are forced to use any of a variety of shelters, not only that in or under a squirrel nest. If so, their behavior places ever more burden on both deep nocturnal torpor and simultaneously the necessity of shivering all the night long.

The cheechako, in Jack London's story "To Build a Fire," died in the Alaskan winter not because he made one big mistake. He was just unlucky and also made tiny errors that accumulated and amplified, until they made all the difference. I conclude that the kinglet similarly, but oppositely, has no magic key for survival in the cold winter world of snow and ice. Those that live there are lucky and do every little thing just right. The odds of surviving the winter are slim, but the gamble, as in the breaking of the buds and the winter hive exits of the bees, has risks that must be taken. The unlucky rolls of the dice that result in individual deaths are absorbed by a high reproductive rate.

Lucky for a kinglet, it does not know the odds stacked against its individual survival. Presumably it could not contemplate its fate, regret about mistakes, or fret over injustice or lost opportunities. It does not worry about the future, or about life and death. Why can we presume this? Because these mental capacities could only compromise, not aid survival. They could not activate the bird to effective action, because there is so little, if anything, it could do to change things in its world where the relevant things—ice storms, a subzero night, winds, food scarcity—are ruled by chance. Undampened enthusiasm and raw drive *would* matter. I do not and cannot ever know the combination of happiness, hunger, or emotions that energize a bird. But whenever I've watched kinglets in their nonstop hopping, hovering, and searching, seen their intimate expressions, and heard their constant chatter of *tsees,* songs, and various calls, I've felt an infectious hyperenthusiasm flow from them, and sensed a grand, boundless zest for life. They could not survive without that in their harsh world. Like us, they are programmed for optimism.

I am gladdened to know that a population of these wraiths of the forest thrives. When I'm in the warmth of my cabin and hear gusts of wind outside that moan through the woods and shake the cabin on wintry nights, I will continue to marvel at and wonder how the little featherpuffs are faring. They defy the odds and the laws of physics, and prove that the fabulous is possible.

01 FIRE AND ICE

Imbrie J., and K. P. Imbrie, 1979. *Ice Ages: Solving the Mystery*. Cambridge, Mass.: Harvard University Press.

Madigan, M. T., and B. L. Marrs. 1997. Extremophiles. *Scientific American* 276(4):82–87.

Margulis, L. 1982. *Early Life*. Boston: Science Books International.

02 SNOW AND THE SUBNIVIAN SPACE

Bentley, W. A., and W. J. Humphreys. 1931. *Snow Crystals*. New York: McGraw-Hill.

Blanchard, D. C. 1998. *The Snowflake Man*. Blacksburg, Va.: McDonald and Woodward Publishers.

Borland, Hal. 1971. A lifetime of snowflakes. *Audubon* 73:59–65.

Burlington Free Press. Bentley's contribution. *Burlington Free Press*. December 24, 1931.

Court, G. 1998. Winter grays. *Natural History* (February): 50–54.

Marchand, P. J. 1993. The underside of winter. *Natural History* (February): 51–56.

Pruitt, W. O., Jr. 1960. *Animals of the North*. New York: Harper and Row.

Thaler, E. 1982. Ornithologisches in Schnee. *Dic Gefiederte Welt* (March): 90–92.

03 A LATE WINTER WALK

Benkman, C. W. 1987. Food profitability and the foraging ecology of crossbills. *Ecological Monographs* 57(3):251–267.

———. 1989. On the evolution and ecology of island populations of crossbills. *Evolution* 43:1324–1330.

———. 1990. Intake rates and the timing of crossbill reproduction. *Auk* 197:376–386.

Bent, A. C. 1968. *Life Histories of North American Cardinals, Grosbeaks, Buntings, Towhees, Finches, Sparrows, and Allies*. Part I. New York: Dover Publications.

Grinnell, J. 1900. Birds of the Kotzebue Sound region. *Pacific Coast Avifauna*, No. 1.

Macoun, J. 1909. *Catalogue of Canadian Birds*, 2nd edition.

Palmer, R. S. 1949. Maine birds. *Bull. Mus. Camp. Zool.* Vol. 102.

Porter, E. 1966. *Summer Island*. New York: Balentine Books. P. 78.

Smith, B. E. 1949. White-winged crossbills nesting in Maine. *Maine Audubon Soc. Bull.* 5:12–13.

Stone, W. 1937. *Bird Studies at Old Cape May*. Vol. II. New York: Dover Publications.

Tufts, H. F. 1906. Nesting of crossbills in Nova Scotia. *Auk* 23:339–340.

04 TRACKING A WEASEL

Sandell, M. 1988. Stop-and-go stoats. *Natural History* (June): 55–64.

Snyder, D. P. 1982. *Tamias striatus*. In *Mammalian Species*, No. 168:1–8. The American Society of Mammalogists.

05 NESTS AND DENS

Frazier, A., and V. Nolan. 1959. Communal roosting by the Eastern Bluebird in winter. *Bird Banding* 30:219–226.

Ghalambor, C. K., and T. E. Martin. 1999. Red-breasted Nuthatch (*Sitta canadensis*). In *The Birds of North America*, edited by A. Poole and F. Gill. No. 459. Philadephia: The Birds of North America, Inc.

Guntert, M., D. Hay, and R. P. Balda. 1988. Communal roosting in the Pygmy Nuthatch: A winter survival strategy. *Proc. Intern. Ornthol. Congr.* 19:1964–1972.

Headstrom, R. 1970. *A Complete Field Guide to Nests in the United States*. New York: Ives Washburn, Inc.

Heinrich, B. 1994. Bald-faced hunters. In *In a Patch of Fireweed*. Cambridge, Mass.: Harvard University Press. Pp. 152–162.

Ingold, J. L., and R. Galati. 1997. Golden-crowned kinglet (*Regulus satrapa*). In *The Birds of North America*, edited by A. Poole and F. Gill. No. 301. Philadelphia: The Birds of North America, Inc.

Knorr, O. A. 1957. Communal roosting of the Pygmy Nuthatch. *Condor* 59:398.

Pitts, T. D. 1976. Fall and winter roosting habits of Carolina chickadees. *Wilson Bull.* 88:603–610.

Rogers, L. 1981. A bear in its lair. *Natural History* 90 (October): 64–70.

Thaler, E. 1990. *Die Goldhähnchen*. Wittenberg Lutherstadt: A. Ziemsen Verlag.

Walsberg, G. E. 1990. Communal roosting in a very small bird: Consequences for the thermal and respiratory gas environments. *Condor* 92:795–798.

White, F. N., G. A. Bartholomew, and J. L. Kinney. 1975. The thermal significance of the nest of the sociable weaver, *Philetairus socius:* winter observations. *Ibis* 117:171–179.

06 FLYING SQUIRRELS IN A HUDDLE

Cowan, I. McT. 1936. Nesting habits of the flying squirrel. *Canadian Field Naturalist* 46:58–60.

DeCoursey, P. J. 1961. Effect of light on the circadian activity rhythms of the flying squirrel, *Glaucomys volans*. *Z. Vergl. Physiol*. 44:331–354.

Dunlap, J. C. 1999. Molecular bases for circadian clocks. *Cell* 96:271–290.

French, A. R. 1977. Circannual rhythmicity and entrainment of surface activity in the hibernator, *Perognathus longimembris*. *J. Mammal*. 58:37–43.

Heinrichs, J. 1983. The winged snail darter. *J. Forestry* 81:212–215, 262.

Maser, C., R. Anderson, and E. N. Bull. 1981. Aggregation and sex segregation in northern flying squirrels in northeastern Oregon, an observation. *Murrelet* 62:54–55.

Maser, C., J. M. Trappe, and R. A. Nausebaum. 1978. Fungalsmall mammal interrelationships with emphasis on Oregon coniferous forests. *Ecology* 59:799–809.

McShea, W. J., and D. M. Madison. 1984. Communal nesting between reproductively active females in a spring population of *Microtus pennsylvanicus*. *Can. J. Zool*. 62:344–346.

Muul, I. 1968. Behavioral and physiological influences on the distributions of the flying squirrels, *Glaucomys volans*. *Misc. Publ. Mus. Zool., Univ. Michigan* 134:1–66.

Osgood, F. L. 1935. Apparent segregation of sexes in flying squirrels. *J. Mammal*. 16:236.

Rust, H. J. 1946. Mammals of northern Idaho. *J. Mammal.* 27:308–327.

Weigl, P. D., and D. W. Osgood. 1974. Study of northern flying squirrel, *Glaucomys sabrinus*, by temperature telemetry. *Amer. Midland Naturalist.* 92:482–486.

Wells-Gosling, N., and L. R. Heaney. 1984. *Glaucomys sabrinus.* In *Mammalian Species,* No. 229:1–8.

Young, M. W. 2000. The tick-tock of the biological clock. *Scientific American* (March): 64–77.

07 HIBERNATING SQUIRRELS (HEATING UP TO DREAM)

Adolph, E. F. 1951. Responses to hypothermia in several species of infant mammals. *Am. J. Physiol.* 166:75–91.

Barnes, B. M. 1989. Freeze avoidance in a mammal: body temperatures below 0°C in an Arctic Hibernator. *Science* 244:1593–1595.

———. 1996. Sang froid. *The Sciences* (September/October): 12–14.

Barnes, B. M., Omtzigt, C., and Daan, S. 1993. Hibernators periodically arouse in order to sleep. In *Life in the Cold: Ecological, Physiological, and Molecular Mechanisms,* edited by C. Carey et al. Boulder, Colo.: Westview Press. Pp. 555–558.

Barnes, B. M., and Ritter, D. 1993. Patterns of body temperature change in hibernating arctic ground squirrels. In *Life in the Cold: Ecological, Physiological, and Molecular Mechanisms,* edited by C. Carey et al. Boulder, Colo.: Westview Press. Pp. 119–130.

Bartholomew, G. A., and T. J. Cade. 1957. Temperature regulation, hibernation, and aestivation in the little pocket mouse, *Perognathus longimembris. J. Mammal.* 38:60–72.

Bartholomew, G. A., and J. W. Hudson. 1960. Aestivation in the Mojave ground squirrel, *Citellus mohavensis. Bull. Mus. Comp. Zool. Harvard* 124:193–208.

Bartholomew, G. A., and R. MacMillen. 1961. Oxygen consumption, estivation and hibernation in the kangaroo mouse, *Microdipodops pallidus*. *Physiol. Zool.* 34:177–183.

Blumer, W. F. C., and L. Cole. 1959. Various degrees of hypothermia in mice. *J. Applied Physiol.* 14:987–989.

Boyer, B. B., and B. M. Barnes. 1999. Molecular and metabolic aspects of mammalian hibernation. *BioScience*. 49(9):713–724.

Buck, C. L., and B. M. Barnes. 1999. Temperature of hibernacula and changes in body composition of Arctic ground squirrels. *J. Mammal.* 80(4):1264–1276.

Cade, T. J. 1963. Observations on torpidity in captive chipmunks of the genus *Eutamias*. *Ecology* 44:255–261.

———. 1964. The evolution of torpidity in rodents. *Annales Academiae Scientiarum Fennicae.* Series A. IV. *Biologica* 71(6):79–112.

Daan, S., B. M. Barnes, and A. M. Strijkstra. 1991. Warming up to sleep? Ground squirrels sleep during arousal from hibernation. *Neurosci. Lett.* 1238:265–268.

Dubois, R. 1896. *Physiologie Compare de la Marmotte.* Paris: Masson.

Folk, E., Jr. 1967. Physiological observations of subarctic bears under winter den conditions. In *Hibernation and Torpor in Mammals and Birds,* edited by C. P. Lyman et al. New York: Academic Press.

Hayward, J. S., and C. P. Lyman. 1967. Nonshivering heat production during arousal from hibernation and evidence for the contribution of brown fat. In *Hibernation and Torpor in Mammals and Birds,* edited by C. P. Lyman et al. New York: Academic Press.

Irving, L., H. Krough, and M. Monson. 1955. The metabolism of some Alaskan animals in winter and summer. *Physiol. Zool.* 28:173–185.

Luyet, B. J. 1964. On the state of water in the tissues of hibernators. *Annales Academiae Scientiarum Fennicae.* Series A. IV. *Biologica* 71(21):298–309.

Lyman, C. P. 1958. Oxygen consumption, body temperature and heart rate of woodchucks entering hibernation. *Amer. J. Physiol.* 194:83–91.

Lyman, C. P. 1982a. Hibernation: Responses to external challenges. In *Hibernation and Torpor in Mammals and Birds,* edited by C. P. Lyman et al. New York: Academic Press. Pp. 176–205.

Lyman, C. P. 1982b. Recent theories of Hibernation. In *Hibernation and Torpor in Mammals and Birds,* edited by C. P. Lyman et al. New York: Academic Press. Pp. 283–301.

Lyman, C. P., and P. O. Chatfield. 1953. Hibernation and cortical electrical activity in the Woodchuck (*Marmota monax*). *Science* 117:533–534.

Lyman, C. P., R. C. O'Brien, G. C. Greene, and E. D. Papafrango. 1981. Hibernation and longevity in the Turkish hamster, *Mesocricetus bradti*. *Science* 212:668–670.

Lyman, C. P., J. S. Willis, A. Malan, and L. C. H. Wang. 1982. *Hibernation and Torpor in Mammals and Birds*. New York: Academic Press.

Malan, A. 1980. Enzyme regulation, metabolic rate and acid-base state in hibernation. In *Animals and Environmental Fitness,* edited by R. Gilles. Oxford, U.K.: Pergamon. Pp. 487–501.

Newman, R. 1967. Metabolism in the eastern chipmunk (*Tamias striatus*) and the Southern flying squirrel (*Glaucomys volans*) during winter and summer. In *Mammalian Hibernation III*. Edinburgh and London: Oliver and Boyd. Pp. 64–67.

Panuska, J. A. 1959. Weight patterns and hibernation in *Tamias striatus*. *J. Mammal.* 40:554–566.

Pauls, W. P. 1978. Behavioural strategies relevant to the energy economy of the red squirrel (*Tamiasciurus hudsonicus*). *Can. J. Zool.* 56:1518–1525.

Pengelley, E. T. 1967. The relation of external conditions to the onset and termination of hibernation and estivation. In

Hibernation and Torpor in Mammals and Birds, edited by C. P. Lyman et al. New York: Academic Press. Pp. 1–29.

Pengelley, E. T., and K. C. Fisher. 1957. Onset and cessation of hibernation under constant temperature and light in the golden-mantled ground squirrels (*Citellus lateralis*). *Nature* (London) 180:1371–1372.

———. 1963. The effect of temperature and photoperiod on the early hibernation behavior of captive golden-mantled ground squirrels, *Citellus lateralis tescorum. Can. J. Zool.* 41:1103–1120.

Smalley, R. L., and R. L. Dryer. 1963. Brown fat: Thermogenic effect during arousal from hibernation in the bat. *Science* 140:1333–1334.

Smith, A. U., J. E. Lovelock, and A. S. Parkers. 1954. Resuscitation of hamsters after supercooling or partial crystallization at body temperatures below 0°C. *Nature* (London) 173:1136–1137.

Smith, A. V. 1959. Survival of mammals at body temperatures above and below 0°C. *Proc. XXI Intern. Congr. Physiol. Sci.* Symposia. Pp. 81–87, Buenos Aires.

Stones, R. C., and J. E. Wiebers. 1965. A review of temperature regulation in bats (*Chiroptera*). *Am. Midl. Natur.* 74:155–167.

———. 1967. Temperature regulation in the little brown bat, *Myotis lucifugus.* In *Hibernation and Torpor in Mammals and Birds,* edited by C. P. Lyman et al. New York: Academic Press.

Storey, K. B. 1997. Metabolic regulation in mammalian hibernation: enzyme and protein adaptations. *Comp. Biochem. Physiol. A Physiol.* 118:1115–1124.

Strumwasser, F., J. J. Gilliam, and J. L. Smith. 1964. Long term studies on individual hibernating animals. *Annales Academiae Scientiarum Fennicae.* Series A. IV. *Biologica* 71(29):401–420.

Travis, J. 1997. Chilled brains. *Science News* 152:364–365.

Tucker, V. A. 1962. Diurnal torpidity in the California pocket mouse. *Science* 136:380–381.

Tucker, V. A. 1966. Diurnal torpor and its relation to food consumption and weight changes in the California pocket mouse, *Perognathus californicus. Ecology* 47:245–252.

08 THE KINGLET'S FEATHERS

Ackerman, J. 1998. Dinosaurs take wing: The Origin of Birds. *National Georgraphic* 194(1):74–99.

Dalton, R. 2000. Feathers fly in Beijing. *Nature* 405:992.

Heinrich, B. 1973. *The Hot-Blooded Insects: Mechanisms and Evolution of Thermoregulation.* Cambridge, Mass.: Harvard University Press.

Ji, Q., P. J. Currie, M. A. Novell, and S. A. Ji. 1998. Two feathered dinosaurs from northeastern China. *Nature* 393:753–761.

Norell, M. 2001. The proof is in the plumage. *Natural History* 7(01):58–63.

Perkins, S. 2001. Ticklish debate: How might the feather have evolved? *Science News* 160:106–108.

Prum, R. O. 2002. Why ornithologists should care about the theropod origin of birds. *Auk* 119(1):1–17.

Schmidt-Nielsen, K. 1972. *How Animals Work.* London: Cambridge University Press.

Xu, X., Z. Zhou, and R. O. Prum. 2001. Branched integumentary structures in sinornithosaurus and the origin of feathers. *Nature* 410:200–204.

09 THE KINGLET'S WINTER FUEL

Bent, A. C. 1964. *Life histories of North American thrushes, kinglets, and their allies. U. S. National Museum Bulletin,* No. 196, New York: Dover Publications.

Galati, R. 1991. *Golden-crowned Kinglets: Treetop Nesters of the North Woods.* Ames, Iowa: Iowa State University Press.

Harrison, J. 1996. In *The Nature Reader,* edited by D. Halpern and D. Frank. Hopewell, N. J.: Ecco Press.

Heinrich, B., and R. Bell. 1995. Winter food of a small insectivorous bird, the Golden-crowned Kinglet. *Wilson Bull.* 107:558–561.

Thaler, E. 1990. *Die Goldhähnchen*. Wittenberg Lutherstadt: A. Ziemsen Verlag.

10 HIBERNATING BIRDS

Austin, G. T., and W. G. Bradley. 1969. Additional responses of the poorwill to low temperatures. *Auk* 86:717–725.

Bartholomew, G. A., T. R. Howell, and T. J. Cade. 1957. Torpidity in the white-throated swift, Anna hummingbird and Poorwill. *Condor* 59:145–155.

Bartholomew, G. A., J. W. Hudson, and T. R. Howell. 1962. Body temperature, oxygen consumption, evaporative water loss, and heart rate in the Poor-will. *Condor* 64:117–125.

Beuchat, C. A., S. B. Chaplin, and M. L. Morton. 1979. Ambient temperature and the daily energetics of two species of hummingbirds, *Calypte anna* and *Selasphorus rufus*. *Physiol. Zool.* 52:280–295.

Blem, C. R. 1975. Geographic variation in wind-loading of the House sparrow. *Wilson Bull.* 87:543–549.

———. 1990. Avian energy storage. *Curr. Ornithol.* 7:59–114.

Blem, C. R., and J. F. Pagels. 1984. Mid-winter lipid reserves of the Golden-crowned kinglet. *Condor* 86:461–492.

Boswell, J. 1927. *Life of Samuel Johnson*. Vol 1. London and New York: Oxford University Press. Pp. 371–372.

Buttemer, W. A., L. B. Astheimer, W. W. Weathers, and A. H. Hayworth. 1987. Energy savings attending winter-nest use by Verdin (*Auriparus flaviceps*). *Auk* 104:531–535.

Calder, W. A., and J. Booser. 1973. Hypothermia of broad-tailed hummingbirds during incubation in nature with ecological correlations. *Science* 180:751–753.

Calder, W. A., and J. R. King. 1974. Thermal and caloric relations of birds. In *Avian Biology*, edited by D. S. Farner and J. R. King. Vol. 4. New York: Academic Press. Pp. 259–413.

Carpenter, F. L. 1974. Torpor in an Andean hummingbird: its ecological significance. *Science* 183:545–547.

Carpenter, F. L., and M. A. Hixon. 1988. A new function for torpor: Fat conservation in a wild migrant hummingbird. *Condor* 90:373–378.

Chaplin, S. B. 1974. Daily energetics of the Black-capped chickadee, *Parus atricapillus*, in winter. *J. Comp. Physiol.* 89:321–330.

———. 1976. The physiology of hypothermia in the Black-capped chickadee, *Parus atricapillus*. *J. Comp. Physiol.* B 112:335–344.

———. 1982. The energetic significance of huddling behavior in Common Bushtits (*Psaltriparus minimus*). *Auk* 99:424–430.

Dawson, W. R., and J. W. Hudson. 1970. Birds. In *Comparative Physiology of Thermoregulation*, edited by G. C. Whittow. Vol. 1. New York: Academic Press. Pp. 223–310.

Dawson, W. R., R. L. Marsh, and M. E. Yacoe. 1993. Metabolic adjustments of small passerine birds for migration and cold. *Am. J. Physiol.* 245:R755–R767.

Ghalambor, C. K., and T. E. Martin. 1999. Red-breasted Nuthatch (*Sitta canadensis*). In *The Birds of North America*, edited by A. Poole and F. Gill. No. 459. Philadelphia: The Birds of North America, Inc.

Gibb, J. 1954. Feeding ecology of tits, with notes on treecreepers and Goldcrests. *Ibis* 96:513–544.

Gilman, M. F. 1902. Notes on the Verdin. *Condor* 4:88–89.

Graber, J. W., and R. R. Graber. 1979. Severe weather and bird populations in southern Illinois. *Wilson Bull.* 91:88–103.

Guntert, M., D. Hay, and R. P. Balda. 1988. Communal roosting in the Pygmy Nuthatch: A winter survival strategy. *Proc. Intern. Ornthol. Congr.* 19:1964–1972.

Hainsworth, F. R., and L. L. Wolf. 1970. Regulation of oxygen consumption and body temperature during torpor in a hummingbird, *Eulampis jugalaris*. *Science* 168:368–369.

Hainsworth, F. R., B. G. Collins, and L. L. Wolf. 1977. The function of torpor in hummingbirds. *Physiol. Zoll.* 50:215–222.

Heinrich, B. 1975. Thermoregulation in bumblebees II. Energetics of warm-up and free flight. *J. Comp. Physiol.* 96:155–166.

———. The physiology of exercise of the bumblebee. *American Scientist* 65:455–465.

Heinrich, B., and G. A. Bartholomew. 1971. An analysis of preflight warm-up in the sphinx moth, *Manduca sexta*. *J. Exp. Biol.* 55:223–239.

Heinrich, B., and T. M. Casey. 1973. Metabolic rate and endothermy in sphinx moths. *J. Comp. Physiol.* 82:195–206.

Heinrich, B., and T. P. Mommsen. 1985. Flight of winter moths near 0°C. *Science* 228:177–179.

Heinrich, B., and R. Bell. 1995. Winter food of a small insectivorous bird, the Golden-crowned kinglet. *Wilson Bull.* 107:558–561.

Hill, R. W., D. L. Beaver, and J. H. Veghte. 1980. Body surface temperatures and thermoregulation in the Black-capped chickadee (*Parus atricapillus*). *Physiol. Zool.* 53:305–321.

Howell, T. R., and G. A. Bartholomew. 1959. Further experiments on torpidity in the Poor-will. *Condor* 61:180–186.

Ingold, J. L., and R. Galati. 1997. Golden-crowned kinglet (*Regulus satrapa*). In *The Birds of North America,* edited by A. Poole and F. Gill. No. 301. Philadelphia: The Birds of North America, Inc.

Jaeger, E. C. 1948. Does the Poor-will "hibernate"? *Condor* 50:45–46.

Jaeger, E. C. 1949. Further observations on the hibernation of the Poor-will. *Condor* 51:105–109.

Kessel, B. 1976. Winter activity patterns of Black-capped Chickadees in interior Alaska. *Wilson Bull.* 88:36–61.

Lasiewski, R. C., and W. R. Dawson. 1964. Physiological responses to temperature in the common nighthawk. *Condor* 66:477–490.

Ligon, J. D. 1970. Still more responses of the Poor-will to low temperature. *Condor* 72:496–498.

Lyman, C. P., J. S. Willis, A. Malan, and L. C. H. Wang. 1982. *Hibernation and Torpor in Mammals and Birds*. New York and London: Academic Press.

Marsh, R. L., and W. R. Dawson. 1982. Substrate metabolism in seasonally acclimated American Goldfinches. *Amer. J. Physiol.* 242:R563–R569.

Marshall, T. T., Jr. 1955. Hibernation in captive goatsuckers. *Condor* 57:129–134.

McAtee, W. L. 1947. Torpidity in birds. *Amer. Midl. Natur.* 38:191–206.

Nakamura, H., and Y. Wako. 1988. Food storing behavior of Willow tit, *Parus montanus. J. Yamashina Inst. Ornith.* 20:721–736.

Peipunen, V. A. 1966. The diurnal heterothermia and torpidity in the Nightjar (*Caprimulgus europaeus*). *Ann. Acad. Sci. Fennicae.* Ser. A. 101:1–35.

Perrins, C. M. 1976. *British Tits*. Glasgow: Collins & Co.

Pitts, T. D. 1976. Fall and winter roosting habits of Carolina chickadees. *Wilson Bull.* 88:603–610.

Reinertsen, R. E., and S. Haftorn. 1986. Different metabolic strategies of northern birds for nocturnal survival. *J. Comp. Physiol.* B 156:655–663.

Reinertsen, R. E., S. Haftorn, and R. Thaler. 1988. Is hypothermia necessary for winter survival of the Goldcrest, *Regulus regulus? J. Ornithol.* 129:433–437.

Rising, J. D., and J. W. Hudson. 1974. Seasonal variation in the metabolism and thyroid activity of the Black-capped chickadee (*Parus atricapillus*). *Condor* 76:198–203.

Smith, Susan M. 1991. *The Black-capped Chickadee: Behavioral Ecology and Natural History*. Ithaca, N.Y.: Cornell University Press.

Steen, J. 1958. Climatic adaptation in some northern birds. *Ecology*. 39:625–629.

Vogt, F. D., and B. Heinrich. 1983. Thoracic temperature variations in the onset of flight in dragonflies (*Odonata: Anisoptera*). *Physiol. Zool.* 56:236–241.

Walsberg, G. E. 1990. Communal roosting in a very small bird: Consequences for the thermal and respiratory gas environments. *Condor* 92:795–798.

Webster, M. C. 1999. Verdin (*Auriparus flaviceps*). In *The Birds of North America*, edited by A. Poole and F. Gill. No. 470, Philadelphia: The Birds of North America, Inc.

Withers, P. C. 1977. Respiration, metabolism, and heat exchange of euthermic and torpid poorwills and hummingbirds. *Physiol. Zool.* 50:43–52.

Wolf, L. L., and F. R. Hainsworth. 1972. Environmental influence on regulated body temperature in torpid hummingbirds. *Comp. Biochem. Physiol.* 41A:167–173.

Zonov, G. B. 1967. On the winter roosting of Paridae in Cicbaikal. *Ornitologiya* 8:351–354. (In Russian; quoted in S. M. Smith.)

11 TORPID TURTLES UNDER ICE

Belkin, D. A. 1968. Aquatic respiration and underwater survival of two freshwater turtle species. *Respir. Physiol.* 4:1–14.

Brown, G. P., and R. J. Brooks. 1994. Characteristics and fidelity to hibernacula in a northern population of snapping turtles, *Chelydra serpentina*. *Copeia* 1994:222–226.

Crocker, C. E., T. E. Graham, G. R. Ultsch, and D. C. Jackson.

2000. Physiology of common map turtles (*Graptemys geographica*) hibernating in the Lamoille River, Vermont. *J. Exp. Zool.* 286:143–148.

Gatten, R. E. 1981. Anaerobic metabolism in freely diving painted turtles (*Chrysemys picta*). *J. Exp. Biol.* 212:377–385.

Graham, T. E., and A. A. Graham. 1992. Metabolism and behavior of wintering common map turtles, *Graptemys geographica*, in Vermont. *Canadian Field Naturalist* 106:517–519.

Graham, T. E., C. B. Graham, C. E. Crocker, and G. R. Ultsch, 2000. Dispersal from and fidelity to a hibernaculum in a northern Vermont population of common map turtles, *Graptemys geographica*. *Canadian Field Naturalist* 114:405–408.

Herbert, C. V., and D. C. Jackson. 1985. Temperature effects on the responses to prolonged submergence in the turtle *Chrysemys picta bellii*. I. Blood acid-base and ionic changes during and following anoxic submergences. *Physiol. Zool.* 58:655–669.

Jackson, D. C. 1968. Metabolic depression and oxygen depletion in the diving turtle. *J. Appl. Physiol.* 24:503–509.

Jackson, D. C., and G. R. Ultsch. 1982. Long-term submergence at 3°C of the turtle, *Chrysemys picta bellii*, in normoxic and severely hypoxic water. II. Extracellular ionic responses to extreme lactic acidosis. *J. Exp. Biol.* 96:29–43.

Pluto, T. G., and E. D. Bellis. 1988. Seasonal and annual movements of riverine map turtles, *Graptemys geographica*. *J. Herpet.* 22:152–158.

Ultsch, G. R. 1985. The viability of nearctic freshwater turtles submerged in anoxia and normoxia at 3 and 10°C. *Comp. Biochem. Physiol.* 81A(3):607–611.

———. 1988. Blood gases, hematocrit, plasma ion concentrations, and acid-base status of musk turtles (*Sternotherus odoratus*) during simulated hibernation. *Physiol. Zool.* 61(1):78–94.

Ultsch, G. R., and D. C. Jackson. 1982. Long-term submergence at 3°C of the turtle, *Chrysemys picta bellii,* in normoxic and severely hypoxic water. I. Survival, gas exchange and acid-base status. *J. Exp. Biol.* 96:11–18.

Ultsch, G. R., and D. Lee. 1983. Radiometric observations of wintering snapping turtles (*Chelydra serpentina*) in Rhode Island. *J. Alabama Acad. Sci.* 54(4):200–206.

Ultsch, G. R., C. V. Herbert, and D. C. Jackson. 1984. The comparative physiology of diving in North American freshwater turtles. I. Submergence tolerance, gas exchange, and acid-base balance. *Physiol. Zool.* 57:620–631.

Ultsch, G. R., R. W. Hanley, and T. R. Bauman. 1985. Responses to anoxia during simulated hibernation in northern and southern painted turtles. *Ecology* 66:388–395.

Ultsch, G. R., M. E. Carwile, C. E. Crocker, and D. C. Jackson. 1999. The physiology of hibernation among painted turtles: The Eastern painted turtle *Chrysemys picta picta. Physiol. Biochem. Zool.* 72(4):493–501.

12 ICED-IN WATER RODENTS

Bazin, R. C., and R. A. MacArthur. 1992. Thermal benefits of huddling in the muskrat (*Ondatra zibethicus*). *J. Mammal.* 73(3):559–564.

Bovet, J., and E. F. Oertli. 1974. Free-running circadian activity rhythms in free-living beaver (*Castor canadensis*). *J. Comp. Physiol.* 92:1–10.

Calder, W. A. 1969. Temperature relations and underwater endurance of the smallest homeothermic diver, the water shrew. *Comp. Biochem. Physiol.* 30:1075–1082.

Heinrich, B. 1984. Strategies of thermoregulation and foraging in two wasps, *Dolichovespula maculate* and *Vespula vulgaris. J. Comp. Physiol.* B154:175–180.

MacArthur, R. A. 1979. Seasonal patterns of body temperature and activity in free-ranging muskrats (*Ondatra zibethicus*). *Can. J. Zool.* 57:25–33.

———. 1984a. Microenvironment gas concentrations and tolerance to hypercapnia in the muskrat (*Ondatra zibethicus*). *Physiol. Zool.* 57(1):85–98.

———. 1984b. Aquatic thermoregulation in the muskrat (*Ondatra zibethicus*): Energy demands of swimming and diving. *Can. J. Zool.* 62:241–248.

———. 1992a. Foraging range and aerobic endurance of muskrats diving under the ice. *J. Mammal.* 73(3):565–569.

———. 1992b. Gas bubble release by muskrats diving under ice: Lost gas or a potential oxygen pool? *J. Zool. Lond.* 226:151–164.

MacArthur, R. A., M. M. Humphries, and D. Jeske. 1997. Huddling behavior and the foraging efficiency of muskrats. *J. Mammal.* 78(3):850–858.

Smith, D. W., R. O. Peterson, T. D. Drummer, and D. S. Sheputis. 1991. Over-wintering activity and body temperature patterns in northern beavers. *Can. J. Zool.* 69:2178–2182.

Smith, D. W., T. D. Drummer, and R. Peterson. 1994. Reply—studies of beaver activity and body temperature in a historical perspective. *Can. J. Zool.* 72:567–569.

13 FROZEN FROGS ON ICE

Bentley, P. J. 1996. Adaptations of Amphibia to arid environments. *Science* 152:619–623.

Burrough, J. 1951. A sharp lookout. In *John Burrough's America*, edited by F. A. Wiley. Devia-Adair Co.

Schmid, W. D. 1965. Some aspects of the water economics of nine species of amphibians. *Ecology* 46:261–269.

———. 1982. Survival of frogs in low temperature. *Science* 215:697–698.

Storey, K. B., and J. M. Storey. 1984. Biochemical adaptation for freezing tolerance in the wood frog, *Rana sylvatica*. *J. Comp. Physiol.* B155:29–36.

———. 1985. Adaptations of metabolism for freeze tolerance in the gray tree frog, *Hyla versicolor*. *Can. J. Zool.* 63:49–54.

———. 1986. Freeze tolerant frogs: Cryoprotectants and tissue metabolism during freeze-thaw cycles. *Can. J. Zool.* 64:49–56.

———. 1987. Persistence of freeze tolerance in terrestrial hibernating frogs after spring emergence. *Copeia* 3:720–726.

———. 1988. Freeze tolerance: constraining forces, adaptive mechanisms. *Can. J. Zool.* 66:1122–1127.

———. 1990. Frozen and alive. *Scientific American* (December): 92–97.

Tester, J. R., and W. J. Breckenridge. 1964. Winter behavior patterns of the Manitoba toad, *Bufo hemiophrys*, in northwestern Minnesota. *Annales Academiae Scientiarum Fennicae*. Series A. IV. *Biologica* 71(31):424–431.

14 INSECTS: FROM THE DIVERSITY TO THE LIMITS

Alonso, M. A., J. I. Glendinning, and L. P. Browers. 1993. The influence of temperature on crawling, shivering, and flying overwintering monarch butterflies in Mexico. In *Biology and Conservation of the Monarch Butterfly*, edited by S. B. Malcolm and M. P. Zalucki. Science Series No. 38. Los Angeles: Natural History Museum of Los Angeles County. Pp. 309–314.

Asahina, E. 1969. Frost resistance in insects. *Advances in Insect Physiology* 6:1–49.

Asahina, E., and K. Tanno. 1964. A large amount of trehalose in a frost resistant insect. *Nature* 204:1222.

Baust, J. G., R. Grandee, G. Condon, and R. E. Morissey. 1979. The diversity of overwintering strategies utilized by separate populations of gall insects. *Physiol. Zool.* 52:572–580.

Baust, J. G., and R. R. Rojas. 1985. Review—Insect cold hardiness: Facts and fancy. *J. Insect Physiol.* 31:755–759.

Diamond, J. M. 1989. Cryobiology: Resurrection of frozen animals. *Nature* 339:509–580.

Discover. 1987. Today a frozen dog, tomorrow the iceman. *Discover* 8 (June): 9.

Duman, J., and K. Howarth. 1983. The role of hemolymph proteins in the cold tolerance of insects. *Ann. Rev. Physiol.* 45:261–270.

Heinrich, B. 1987. Thermoregulation by winter-flying endetheric moths. *J. Exp. Biol.* 127:313–332.

Heinrich, B., and T. P. Mommsen. 1985. Flight of winter moths near 0°C. *Science* 228:177–179.

Hinton, H. E. 1960. A fly larva that tolerates dehydration and temperatures of minus 270°C to +102°C. *Nature* 188:333–337.

Holland, W. J. 1968. *The Moth Book*. New York: Dover Publications.

Howarth, K. L., and J. G. Duman. 1984. Yearly variations in overwintering mechanisms of the cold-hardy beetle, *Denroides canadensis. Physiol. Zool.* 57(1):40–45.

Kukal, O. 1988. Caterpillars on ice. *Natural History* (January): 36–40.

Kukal, O., B. Heinrich, and J. Duman. 1988. Behavioral thermoregulation in the freeze-tolerant Arctic caterpillar, *Gynaephora groenlandica. J. Exp. Biol.* 138:181–193.

Leader, J. P. 1962. Tolerance to freezing of hydrated and partially hydrated larvae of *Polypedilum* (Chironomidae). *J. Insect Physiol.* 8:155–163.

Lee, R. E., Jr. 1989. Insect cold-hardiness: To freeze or not to freeze. *Bioscience* 39(5):308–313.

Lee, R. E., Jr., and D. L. Denlingers, eds. 1991. *Insects at Low Temperature*. New York: Chapman and Hall.

Mansingh, A., and B. N. Smallman. 1972. Variation in polyhydric alcohol in relation to diapause and cold-hardiness in the larva of *Isia isabella*. *J. Insect Physiol.* 18:1565–1576.

Mazur, P. 1984. Freezing of living cells: mechanisms and implications. *Amer. J. Physiol.* 247:C125–C142.

Miller, L. K. 1969. Freezing tolerance in an adult insect. *Science* 166:105–106.

Myers, M. T. 1985. A southward return migration of Painted Lady butterflies, *Vanessa cardui*, over southern Alberta in the fall of 1983, and biometeorological aspects of their outbreaks into North America and Europe. *Canadian Field-Naturalist* 99:147–155.

Salt, R. W. 1959. Survival of frozen fat body cells in an insect. *Nature* 184:1426.

———. 1961. Principles of insect cold hardiness. *Ann. Rev. Entomol.* 6:58–74.

Schmid, W. D. 1982. Survival of frogs in low temperature. *Science* 215:697–698.

Storey, K. B., and J. M. Storey. 1983. Biochemistry of freeze tolerance in terrestrial insects. *Trends in Biochemical Sciences* 8(7):242–245.

———. 1984. Biochemical adaptation for freezing tolerance in the wood frog, *Rana sylvatica*. *J. Comp. Physiol.* B 155:29–36.

———. 1985. Adaptations of metabolism for freeze tolerance in the gray tree frog, *Hyla versicolor*. *Can. J. Zool.* 63:49–54.

———. 1987. Persistence of freeze tolerance in terrestrial hibernating frogs after spring emergence. *Copeia* 1987(3):720–726.

———. 1988. Freeze tolerance in animals. *Physiol. Rev.* 68:27–88.

———. 1990. Frozen and alive. *Scientific American* (December): 92–96.

Tanno, K. 1968. Frost resistance in the poplar sawfly, *Trichiocampus populi* V. Freezing injury at the liquid nitrogen temperature. *Low Temp. Sci. Ser.* B 26:76–84.

Vogel, S. 1998. Cold storage. *Discover* (February): 52–54.

REFERENCES

Walton, R. K., and L. P. Brower. 1996. Monitoring the fall migration of the monarch butterfly *Danaus plexippus* L. (Nymphalidae: Danaidae) in eastern North America: 1991–1994. *J. Lepidopterists' Society* 50:1–20.

Zachariassen, K. E., and H. T. Hamel. 1976. Nucleating agents in the hemolymph of insects tolerant to freezing. *Nature* (London) 262:285–287.

15 MICE IN WINTER

Choate, J. R. 1973. Identification and recent distribution of white-footed mice (*Peromyscus*) in New England. *J. Mammal.* 54:41–49.

Glaser, H., and S. Lustick. 1975. Energetics and nesting behavior of the northern white-footed mouse, *Peromyscus leucopus noveboracensis. Physiol. Zool.* 48:105–113.

Hamilton, W. J., Jr. 1935. Habits of jumping mice. *Amer. Midl. Natur.* 16:187–200.

Nestler, J. R. 1990. Relationship between respiratory quotient and metabolic rate during entry to and arousal from daily torpor in deer mice (*Peromyscus maniculatus*). *Physiol. Zool.* 63(3):504–515.

Nicholson, A. J. 1937. A hibernating jumping mouse. *J. Mammal.* 18:103.

Parren, S. G., and D. E. Capen. 1985. Local distribution and coexistence of two species of *Peromyscus* in Vermont. *J. Mammal.* 66(1):36–44.

Pierce, S. S., and F. D. Vogt. 1993. Winter acclimatization in *Peromyscus maniculatus gracilis, P. leucopus noveboracencis*, and *P. l. leucopus. J. Mammal.* 74(3):665–677.

Sealander, A. 1962. Seasonal changes in blood values of deer mice and other small mammals. *Ecology* 43(1):107–119.

Sealander, J. A. 1951. Survival of *Peromyscus* in relation to environmental temperature and acclimation at high and low temperatures. *Amer. Midl. Natur.* 46:257–311.

———. 1952. The relationship of nest protection and huddling to survival of *Peromyscus* at low temperature. *Ecology* 33:63–71.

Sheldon, C. 1938a. Vermont jumping mice of the genus *Zapus*. *J. Mammal.* 19:324–332.

———. 1938b. Vermont jumping mice of the genus *Napaeozapus*. *J. Mammal.* 19:444–453.

Stupka, A. 1934. Woodland jumping mice. Nature notes from *Acadia* 3:6.

Vogt, F. D., and G. R. Lynch. 1982. Influence of ambient temperature, nest availability, huddling, and daily torpor on energy expenditure in the white-footed mouse *Peromyscus leucopus*. *Physiol. Zool.* 55(1):56–63.

Walton, M. A. 1903. *A Hermit's Wild Friends or Eighteen Years in the Woods.* Boston: Dana Estes Co.

16 SUPERCOOL(ED) HOUSEGUESTS (WITH AND WITHOUT ANTIFREEZE)

Barnes, B. M., J. L. Barger, J. Seares, P. C. Tacguard, and G. L. Zuercher. 1996. Overwintering in yellowjacket queens (*Vespula vulgaris*) and green stinkbugs (*Elasmostethus interstinctus*) in subarctic Alaska. *Physiol. Zool.* 69(6):1469–1480.

Duman, J. G., and J. L. Patterson. 1978. The role of ice nucleators in the frost tolerance of overwintering queens of the bald-faced hornet. *Comp. Biochem. Physiol.* 59A:69–72.

Duman, J. G., D. W. Wu, L. Xu, D. Tursman, and T. M. Olson. 1991. Adaptations of insects to subzero temperatures. *Quart. Rev. Biol.* 66:387–410.

Barclay, R. M. R. 1982. Night roosting behavior of the little brown bat, *Myotis lucifugus*. *J. Mammal.* 63:464–474.

Brower, L. P., and S. B. Malcolm. 1991. Animal migration: Endangered phenomena. *Amer. Zool.* 31:265–276.

Brower, L. P., W. H. Calvert, L. E. Hedrick, and J. Christian. 1977. Biological observations on an overwintering colony of monarch butterflies (*Danaus plexippus*, Danaidae) in Mexico. *J. Lepid. Soc.* 31(4):232–242.

Calvert, W. H., L. E. Hedrick, and L. P. Brower. 1979. Mortality of the monarch butterfly (*Danaus plexippus* L.): Avian predation at five overwintering sites in Mexico. *Science* 204:848–851.

Flyger, V. F. 1969. The 1968 squirrel "migration" in the eastern United States. *Trans. Northeast Sect. Wildlife Society* 26:69–70.

Henshaw, R. E., and G. E. Folk, Jr. 1966. Relation of thermoregulation to seasonally changing microclimate in two species of bats (*Myotis lucifugus* and *M. sodalis*). *Physiol. Zool.* 39:223–236.

Hock, R. J. 1951. The metabolic rates and body temperatures of bats. *Biol. Bull.* 101:289–299.

Humphrey, S. R., and J. B. Cope. 1977. Survival rates of the endangered Indiana bat, *Myotis sodalis*. *J. Mammal.* 58:32–36.

Kammer, A. E. 1970. Thoracic temperature, shivering, and flight in the monarch butterfly, *Danaus plexippus* (L.). *Z. Vergl. Physiol.* 68:334–344.

Masters, A. R., S. B. Malcolm, and L. P. Brower. 1988. Monarch butterfly (*Danaus plexippus*) thermoregulatory behavior and adaptations for overwintering in Mexico. *Ecology* 69(2):458–467.

McNab, B. K. 1974. The behavior of temperate cave bats in a subtropical environment. *Ecology* 55:943–958.

Richter, A. R., S. R. Humphrey, J. B. Cope, and V. Brack, Jr. 1993.

Modified cave entrances: Thermal effect on body mass and resulting decline of endangered Indiana bats (*Myotis sodalis*). *Conservation Biology* 7(2):407–415.

Speakman, J. R., and P. A. Racey. 1989. Hibernal ecology of the pipistrelle bat: Energy expenditure, water requirements and mass loss, implications for survival and the function of winter emergence flights. *J. Animal Ecology* 58:797–813.

Tuttle, M. D. 1979. Status, causes of decline and management of endangered gray bats. *J. Wildlife Management* 43:1–17.

Urquhart, F. A. 1976. Found at last: The monarch's winter home. *National Geographic Magazine* 150:160–173.

———. 1987. *Monarch Butterfly, the International Traveler*. Chicago: Nelson-Hall.

Urquhart, F. A., and N. R. Urquhart. 1976. The overwintering site of the eastern population of the monarch butterfly (*Danaus plexippus*; Danaidae) in southern Mexico. *J. Lepid. Soc.* 30:153–158.

18 AGGREGATING FOR WINTER

Gorenzel, P. W., and T. P. Salmon. 1995. Characteristics of American crow urban roosts in California. *J. Wildlife Management* 59(4):638–645.

Hanson, H. G. 1946. Crow centers of the United States. *Oklahoma Game and Fish News* 2(3):4–7, 18.

Heinrich, B. 1989. Communal roosts. In *Ravens in Winter*. New York: Simon and Schuster. Pp. 159–165.

Marzluff, J. M., B Heinrich, and C. S. Marzluff. 1996. Roosts are mobile information centers. *Animal Behav.* 51:89–103.

Shine, R., and R. Mason. 2001. Serpentine cross-dressers. *Natural History* (February): 56–61.

Stouffer, P. C., and D. F. Caccamise. 1991. Roosting and diurnal movements of radio-tagged American crows. *Wilson Bull.* 103(3):387–400.

19 WINTER FLOCKS

Morse, D. H. 1977. Feeding behavior and predator avoidance in heterospecific groups. *BioScience* 27(5):332–339.

Thaler, E. 1990. *Die Goldhähnchen*. Wittenberg Lutherstadt: A. Ziemsen Verlag.

20 BERRIES PRESERVED

Karasov, W. H. 1993. In the belly of the bird. *Natural History* (November): 32–37.

Stiles, E. W. 1984. Fruit for all seasons. *Natural History* (August): 43–53.

21 BEARS IN WINTER

Alt, G. L., and J. M. Grutladavria. 1984. Reuse of black bear dens in northeastern Pennsylvania. *J. Wildlife Management* 48:236–239.

Harlow, H. J., T. Lohuis, T. D. I. Beck, and P. A. Iaizzo. 2001. Muscle strength in overwintering bears. *Nature* 409:997.

Johnson, R. B. 1998. The bearable lightness of being: Bones, muscles, and spaceflight. *Anatomical Record* 253(1):24–27.

Jones, J. D., P. Burnett, and P. Zollman. 1999. The glyoxylate cycle: Does it function in the dormant or active bear? *Compar. Biochem. and Physiol.* B 124:177–179.

LeBlanc, L. Chen, L. Shackelford, V. Sinitsyn, H. Evans, O. Belichenko, B. Schenkman, I. Kozlouskaya, V. Oganov, A. Bakulin, T. Hedrick, and D. Feeback. 2000. Muscle volume, MRI relaxation times (T2), and body composition after spaceflight. *J. Applied Physiol.* 89(6):2158–2164.

Miller, M. K. 1995. Space makes strange bedfellows. *The Sciences* 35(3):12–16.

Nelson, R. A., T. D. I. Beck, and D. L. Steiger. 1984. Ration of serum urea to serum creatine in wild black bears. *Science* 226:841–842.

Ormseth, O. A., M. Nicolson, M. A. Pellymounter, and B. B. Boyer. 1996. Leptin inhibits prehibernation hyperphagia and reduces body weight in arctic ground squirrels. *Amer. J. Physiol.* 271(6):1775.

Rerkin, A. C. 1989. Sleeping beauties. *Discover* (April): 62–65.

Rogers, L. 1981. A bear in its lair. *Natural History* 90 (October): 64–70.

Tietje, W. D., and R. L. Ruff. 1980. Denning behavior of black bears in boreal forest of Alberta. *J. Wildlife Management* 44(4):858–870.

Tinker, D. B., H. J. Harlow, and T. D. I. Beck. 1998. Protein use and muscle-fiber changes in free-ranging, hibernating black bears. *Physiol. Zool.* 71:414–424.

White, R. J., and M. Avener. 2001. Humans in space. *Nature* (London) 409:1115–1118.

22 STORING FOOD

Addison, E. M., R. D. Strickland, and D. J. H. Fraser. 1989. Gray Jays, *Perisoreus canadensis,* and Common Ravens, *Corvus corax,* as predators of winter ticks, *Dermacentor albipictus*. *Can. Field Nat.* 103(3):406–408.

Clayton, N. S., and D. W. Lee. 1998. Memory and the hippocampus in food-storing birds. In *Animal Cognition in Nature,* edited by R. P. Balda, I. M. Pepperberg, and A. C. Kamil. San Diego and London: Academic Press. Pp. 99–118.

Clayton, N. S., and J. R. Krebs. 1994. Hippocampal growth and attrition in birds affected by experience. *Proc. Nat. Acad. Sci. USA* 91:7410–7414.

Dow, D. D. 1965. The role of saliva in food storage by the Gray Jay. *Auk* 82:139–154.

Heinrich, B. 1988. Winter foraging at carcasses by three sympatric corvids, with emphasis on recruitment by the raven, *Corvus corax. Behav. Ecol. and Sociobiol.* 23:141–156.

———. 1999. *Mind of the Raven.* New York: HarperCollins.

Heinrich, B., and J. Pepper. 1998. Influence of competitors on caching behavior in the Common Raven, *Corvus corax. Anim. Behav.* 56:1083–1090.

Källander, H., and H. D. Smith. 1990. Food storing in birds: An evolutionary perspective. In *Current Ornithology*, edited by D. M. Powers. Vol. 7. New York and London: Plenum Press. Pp. 147–207.

Marzluff, J. M., B. Heinrich, and C. S. Marzluff. 1996. Roosts are mobile information centers. *Anim. Behav.* 51:89–103.

Stahler, D. R., B. Heinrich, and D. W. Smith. 2002. The raven's behavioral association with wolves. *Anim. Behav.* (in press).

Strickland, D. 1991. Juvenile dispersal in Gray Jays: Dominant brood member expels siblings from natal territory. *Can. J. Zool.* 69:2935–2945.

Strickland, D., and H. Ouellet. 1993. Gray Jay. In *The Birds of North America*, edited by A. Poole and F. Gill. No. 40. Philadelphia: Academy of Natural Sciences. Washington, D.C.: American Ornithologists' Union.

Vander Wall, S. B., and R. P. Balda. 1981. Ecology and evolution of food-storage behavior in conifer seed-caching corvids. *Z. Tierpsychol.* 56:217–242.

———. 1983. Rememberance of seeds stashed. *Natural History* 92:60–65.

Vander Wall, S. B., and H. E. Hutchins. 1983. Dependence of Clark's Nutcracker, *Nucifraga columbiana*, on conifer seeds during the postfledging period. *Can. Field Nat.* 97:208–214.

Waite, T. A., and D. Strickland. 1997. Cooperative breeding in Gray jays: Philopatric offspring provision juvenile siblings. *Condor* 99:523–525.

23 BEES' WINTER GAMBLE

Frisch, K. von. 1967. *The Dance Language and Orientation of Bees*. Cambridge, Mass.: Harvard University Press.

Heinrich, B. 1979. Thermoregulation of African and European honeybees during foraging, attack, and hive exits and returns. *J. Exp. Biol.* 80:217–229.

———. 1981. The mechanisms and energetics of honeybee swarm temperature regulation. *J. Exp. Biol.* 91:25–55.

Lindauer, M. 1954. Temperaturreguliering und Wasserhaushalt im Bienenstaat. *Z. Vergl. Physiol.* 34:299–345.

———. 1955. Schwarmbienen auf Wohnungsuche. *Z. Vergl. Physiol.* 37:263–324.

Seeley, T. D. 1985. *Honeybee Ecology: A Study of Adaptation in Social Life*. Princeton, N.J.: Princeton University Press.

Seeley, T. D., and R. A. Morse. 1982. How do honeybees find a home? *Scientific American* 247 (October): 158–168.

Seeley, T. D., and J. Tautz. 2001. Worker piping in honeybee swarms and its role in preparing for liftoff. *J. Comp. Physiol.* A 187:667–676.

24 WINTER BUDS

Heinrich, B. 1996. When the bough bends. *Natural History* 2(96):56–57.

Bent, A. C. 1964. *Life Histories of North American Thrushes, Kinglets and Their Allies. U.S. Nat. Mus. Bull.* No. 196, New York: Dover Publications.

Blem, C. R., and J. F. Pagels. 1984. Mid-winter lipid reserves of the golden-crowned kinglet. *Condor* 86:461–492.

Brewster, W. 1888. Breeding of the golden-crowned kinglet (*Regulus satrapa*) in Worcester County, Massachusetts, with a description of its nest and eggs. *Auk* 5:337–344.

Brockie, K. 1984. *One Man's Island.* New York: Harper and Row.

DeGraaf, R. M., and M. Yamasaki. 2001. *New England Wildlife.* Hanover and London: University Press of New England.

Desfayes, M. 1965. Biosystematics note of the genus *Regulus*. *Ardea* 53:82.

Ehrlich, P. R., D. S. Dobkin, and D. Wheye. 1992. *Birds in Jeopardy.* Stanford, Ca.: Stanford University Press.

Galati, R. 1991. *Golden-crowned Kinglets: Treetop Nesters of the North Woods.* Ames, Iowa: Iowa State University Press.

Galati, R., and C. Galati. 1985. Breeding of the golden-crowned kinglet in northern Minnesota. *J. Field Ornithol.* 56:28–40.

Graber, J. W., and R. R. Graber. 1979. Severe weather and bird populations in southern Illinois. *Wilson Bull.* 91(1): 88–103.

Graber, R. R., and J. W. Graber. 1963. A comparative study of bird populations in Illinois, 1906–1909 and 1956–1958. *Ill. Nat. His. Surv. Bull.* 28:283–528.

Gstader, W. 1973. Jahresdynamik der Avifauna des Südwestlichen Innsbrucker Mittelgebirges. *Monticola* 3:1–68.

Heinrich, B. 1992. *The Hot-Blooded Insects.* Cambridge, Mass.: Harvard University Press.

Heinrich, B., and R. Bell. 1995. Winter food of a small insectivorous bird, the Golden-crowned Kinglet. *Wilson Bull.* 107:558–561.

Hilden, O. 1982. Winter ecology and partial migration of the Goldcrest, *Regulus regulus* in Finland. *Ornis Fenn.* 59:99–122.

Hogstad, O. 1984. Variation in numbers, territoriality and flock size of a Goldcrest, *Regulus regulus* population in winter. *Ibis* 126:296–306.

Ingold, J. L., L. A. Weight, and S. I. Guttman. 1988. Genetic differentiation between North America kinglets and comparisons with three allied passerines. *Auk* 105:386–390.

Ingold, J. L., and R. Galati. 1997. Golden-crowned kinglet (*Regulus satrapa*). In *The Birds of North America*, edited by A. Poole and F. Gill. No. 301. Philadelphia: Academy of Natural Sciences. Washington, D.C.: American Ornithologists' Union.

Kania, W. 1983. Preliminary remarks on the migration of North European Goldcrests, *Regulus regulus*. *Ornis. Fenn.*, Suppl. 3:19–20.

Kubisz, M. A. 1989. Burdock as a hazard to Golden-crowned kinglets and other small birds. *Ont. Birds* 7:112–117.

Lagerström, M. 1979. Goldcrests (*Regulus regulus*) roosting in the snow. *Ornis. Fenn.* 56:170–171.

Larrison, E. J., and K. G. Sonnenberg. 1968. Washington birds. *J. Seattle Audubon Soc.* (Seattle, Wash.).

Lepthien, L. W., and C. E. Bock. 1976. Winter abundance of North American kinglets. *Wilson Bull.* 88:482–485.

Löhrl, H. 1955. Schlafgewohnheiten der Baumläufer (*Certhia brachydactyle, C. familiaris*) und arderer Kleinvögel in Kalten Winternächten. *Vogelwarte* 18:71–77.

Palmgren, P. 1932. Zur Biologie von *Regulus r. regulus* (L.) und *Parus atricapillus. Acta Zoologica Fennica* 14:1–113.

———. 1936. Über den Massenwechsel bei *Regulus r. regulus* (L.). *Ornis Fenn.* 13:159–164.

Reinertsen, R. E., S. Haftorn, and E. Thaler. 1988. Is hypothermia necessary for the winter survival of the Goldcrest *Regulus regulus? J. Ornithol.* 4:433–437.

Sabo, S. R. 1980. Niche and habitat relations in subalpine bird communities of the White Mountains of New Hampshire. *Ecol. Monogr.* 50:241–259.

Sheldon, F. H., and F. B. Gill. 1996. A reconsideration of songbird phylogeny, with emphasis on the evolution of titmice and their sylvioid relatives. *Syst. Biol.* 45:473–495.

Sibley, C. G., and J. E. Ahlquist. 1985. The phylogery and classification of the passerine birds, based on comparisons of the genetic material DNA. *Proc. 18th Int. Ornith. Congr. 1984.*

Thaler, E. 1990. *Die Goldhähnchen.* Wittenberg Lutherstadt: A. Ziemsen Verlag.

Thaler, E., and K. Thaler. 1982. Feeding biology of Goldcrests and Firecrests and their segregation by choice of food. *Ökol. Vögel* 4:191–204.

Thaler-Kottek, E. 1986. The genus *Regulus* as an example of different survival strategies: Adaptation to habitat and ethological differentiation. In *Acta XIX Intern. Congr. Ornithol.* II (Ottawa). Pp. 2007–2020.

Page numbers in *italics* indicate illustrations.

About the author

About the book

Read on

Insights,
Interviews
& More...

Meet Bernd Heinrich

© 2007 by Rachel Smolker

BERND HEINRICH is the author of numerous books, including *The Trees in My Forest, A Year in the Maine Woods, Ravens in Winter, One Man's Owl, Bumblebee Economics* (nominated for the National Book Award), and *Mind of the Raven* (winner of the John Burroughs Medal for Natural History Writing). He is professor emeritus of biology at the University of Vermont. Heinrich divides his time between Vermont and the forests of western Maine. ∽

Mementos
A Life in the Field

A letter I wrote at age seven to my father. The letter, in German, was sloppily composed on uneven lines that I had drawn on a piece of blank paper. About half of the words were misspelled. It said: "Dear Papa, will you bring with you the hazel mouse and bring it with you alive and keep it and I also have a bat and it is also alive in a caterpillar cage and I feed it and I have also caught a moorhen for you like the one I caught before it is skinned already and I send greetings Papa and also Mamusha." *(from* The Snoring Bird: My Family's Journey through a Century of Biology, *Ecco/HarperCollins, 2007)*

Mementos *(continued)*

Possibly my first drawing—a watercolor drawing of a "Hirschkäfer" (stag beetle) and an oak leaf. Maybe the oak leaf represents habitat—the tree where it might be found. I don't recall ever seeing one of these beetles, though. It was one I never got for my beetle collection. It must have been drawn by copying from a picture. It was a mythical creature to me. (Curiously, the day after my stressful PhD qualifying exam at UCLA I again sat down and drew a stag beetle much like this one. I think I was reaching back to my past, where so much started.) Papa had a young man (Rolf Grantsau) at the cabin then illustrating ichneumon wasps, and he must have lent me his paint. (Rolf later studied biology at Kiel University, and emigrated to Brazil, where I located him while writing this book. He had followed his mentor—a tropical biologist—there.)

This picture of me was taken a year and several months after we arrived in Maine and my parents bought a rundown farm. I had the freedom of the woods, and soon found a crow's nest and here have a just-fledged new "Jacob," who is begging for food from me. I had put up bird boxes that were occupied by many birds, including kestrels, and I took one of these hawk babies and raised it as well, here holding out to it probably a piece of mouse meat that came from riding on the tractor with Phil Potter as he was mowing the field. (I would jump off periodically to catch bird food.) Both birds were soon independent and on their own. Here I had just come back from a swimming break at a muddy "beach" at our pond about a half mile down the road, at the edge of which I had climbed up into a fir tree to get the baby crow a couple of weeks earlier. Given the feathers of the birds, this picture was taken in July. The barrel in back of my left hand up against the house was a platform I had been standing on to paint the window frames (red and green, I think).

Mementos *(continued)*

Papa and Mamusha went to Tanganyika for over a year in 1964, and I took a break from my studies at the University of Maine and joined them for fourteen months. Our task was collecting birds for the Peabody Museum of Natural History at Yale. Here we have set up a typical campsite a day's march up into the Uluguru Mountains near Morogoro. This was the main work tent. I slept in a pup tent big enough for a mattress and "lived" in the forest hunting birds half the day and skinning and stuffing birds in the work tent in the afternoons and evenings. One day a young boy brought us a baby owl that he had shot with a slingshot. (It had a bloody eye.) It immediately became our pet, and by the time we left there a month or more later it was independent and had left. The owl is here perched on two wooden boxes, of the type Papa had made to have layers of cotton on which we stored the insects we brought back. To my left and Papa's right are packing crates for our food (mostly oatmeal and dried milk).

The Bird Behind the Book

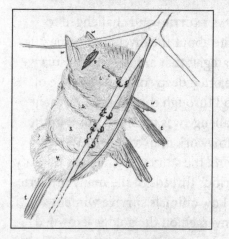

I was very lucky to get a picture showing at least four kinglets huddling up in the branches of a tree on a winter night. The view was looking directly up to the twig they were perching on (although I also got a side view which showed that they all had their heads tucked out of view into their back feathers). Unfortunately, most of the picture was out of focus. However, four birds

The Bird Behind the Book *(continued)*

*could be positively identified from the tight
cluster, mostly by counting tails sticking
out in several directions. Kinglet number
one (the one with its belly feathers in best
focus) had its tail pointing directly toward
the camera lens. Part of its right leg is visible
and its left leg is apparently retracted into
its belly feathers. Kinglets number two
and four also appear to be standing on one
leg. (Knowing that the feet have three toes
pointing forward and one back helps in
determining which foot belongs to whom,
since the tail directions indicate perching
orientation.) In my interpretive sketch of the
photograph, w = wing, t = tail, and l = leg.*

I WAS EXCITED and challenged to
write about the diversity of amazing
strategies that animals use to survive
the often deep cold and absence of
food through the winter. The main
challenge was trying to put it in the
framework of a "story." In *Winter
World* the story that emerged to unify
various threads of the many mysteries
of how animals survive winter
converged on the golden-crowned
kinglet, a bird that is unique in
many ways. It lives in the dense, dark
northern coniferous forests, where
it is difficult to find and few people
ever see it. It has a fluffy, khaki-green
plumage, but from up close one
may on occasion see a crimson crest

framed in bright gold yellow, which it can erect like a flame from the top of its head. It is rare and secretive as well. However, I knew it as a regular resident near my camp in the Maine woods, from which much of my fascination for the winter woods derives. I had for years seen this mite of a bird busily foraging, even after long bone-chilling winter nights when temperatures had dipped to minus 30°F and when the night wind had whipped the trees. I could not imagine how it survived even an hour, much less a night. A winter would be to this bird, who rarely lives a whole year, almost a lifetime. Yet come spring I would find kinglets singing their sweet refrain in a spruce thicket, having survived the winter, and once again getting ready to build their nests.

The kinglet has a tiny, pointed bill, which indicates that it is an insectivore. However, insects hibernate all winter under bark or under the blanket of snow and leaves on the ground, or as larvae encased in solid wood. Knowledge of the kinglets' food base was a central ▶

issue to understand its survival, because almost all insect-eating birds migrate in the fall (some woodpeckers excepted). Nevertheless, kinglets stay active and their very small size indicates that they would very rapidly lose their body heat to the air, and they would have to take in food calories at a much higher rate per body size than other birds.

Kinglets are not much larger than some of the insects I had studied in my previous work as an insect physiologist. I had worked in deciphering the economics of energy balance in small warm-blooded animals, and I became fascinated with how kinglets could compensate for their phenomenally fast heat loss to keep warm at much lower temperatures than those faced by all warm-bodied insects and by most birds. My calculations showed that they should use up all of their food reserves several times every night, to die of starvation several times over. It seemed virtually impossible that they would survive an hour or two at night, much less a sixteen-hour northern night.

> 66 Kinglets are not much larger than some of the insects I had studied in my previous work as an insect physiologist. 99

Kinglets encapsulate the problem of winter survival, and I presumed that they must have mechanisms that, as warm-blooded animals, would likely be similar to ours. I had on many occasions spent hours perched up in a tree in November woods at temperatures near 30°F. And although I was dressed in many layers of clothing and stoked with hot coffee and candy bars after a breakfast of bacon, eggs, and toast, I was unbearably cold after a couple of hours even at those balmy temperatures. To think that a bird weighing 14,165 times less under a tiny layer of feathers could last a sixteen-hour night at temperatures 60°F lower seemed inconceivable to me; they had to do *all* the right things, and maybe more, and do them well. They became iconic to me of winter survival.

The first obvious question was: What could they possibly eat? Whenever I watched them I could see them pick at the branches almost constantly but I had no idea what they were finding. By beating the trees and seeing what fell out onto ▶

the snow, and by making stomach analyses, I found out with great pride and satisfaction that they were feeding mainly on tiny inch-worm caterpillars that spend the winter frozen and silked onto branches. This itself was a discovery, because although the caterpillars were well known, it was assumed that they hibernate under the snow and leaves on the ground. But there was more to learn. Caterpillars are indeed fine food, but they are not a magic solution to surviving freezing at minus 30°F. They explained nothing about how kinglets could survive a winter night.

The next obvious question was where the kinglets spent the night. There had been a report of a kinglet that had apparently entered a squirrel nest. In my mind the "apparently" loomed large because it seemed possible to me that the bird could have snuggled up against the nest to use it as a windbreak. Squirrels are voracious predators of baby birds and eggs, and they would not hesitate to devour a kinglet, if given a chance. I spent countless hours trying to follow

66 I found out with great pride and satisfaction that they were feeding mainly on tiny frozen inch-worm caterpillars that spend the winter frozen and silked onto branches. 99

kinglets through the winter woods, and almost invariably they foraged until it got too dark to follow them. I never once saw them enter a squirrel nest, although I once saw one disappear near one.

Captive kinglets of a closely related European species had been observed to huddle at night. The kinglets I had met in the Maine winter woods traveled all day in groups of from two to six individuals, and it seemed possible that such a group could then immediately huddle and conserve heat (and energy) literally almost "on the spot" where they ended up. They might not have time to search for, or have the luck to find, a warm place to spend the night, but staying together would assure them of an instant huddling partner. Such a strategy would help them to forage for possibly the critical last bit of energy needed to survive the night. However, as logical as this solution might be, and as much as I was convinced of it, it had not been documented. In *Winter World* I could only speculate that the kinglets might compensate for their small size and consequent ▶

energy problem of overnighting in the winter by huddling as a group, but made a sketch of what I thought such a huddle might look like.

Several months after the book was published I happened to see several kinglets at dusk, and as I idly followed them by eye I noticed them flying into a pine tree. More significantly, I did not see them fly out. Their contact calls ceased, and I then realized my amazing opportunity to make a discovery. I came back an hour later with a flashlight, and luckily—since this tree happened to be less than twenty feet tall— I could survey the branches with my flashlight. What a thrill to find the anticipated kinglet huddle. There were four birds with their front ends pressed together into one fluffy ball and with their tails pointed out in three directions.

Later that night I carefully climbed up to examine the kinglet huddle from up close. I needed a photograph for documentation, and had brought my camera. But then when I was finally under them, I slightly jiggled their branch and a head shot out

66 Several months after the book was published I happened to see several kinglets at dusk, and as I idly followed them by eye I noticed them flying into a pine tree. 99

14

from under the back feathers of one bird. It made a little peep. I looked into its tiny dark eyes, expecting the whole flock to dissolve and disperse in an instant. Instead, the bird tucked its head right back into the feathers, and I could follow the rapid vibrations of its breathing movements. I knew then that the birds were not in torpor (from low body temperature). I snapped my picture and the flash did not disturb them. I came back to the birds once more early in the morning before it was light, and they were still in the same place and in the same position, and their tails were still pointing in the same directions as before. They did not come back in the following nights. Presumably they do indeed go to sleep where they end up at the end of the day, which makes sense when they have no time to waste looking for the right place. For them the right place is next to each other.

My photograph, taken shooting directly up to the branch they perched on, is out of focus except for the feet and belly feathers of the first bird in the four-pack, whose tail ▶

The Bird Behind the Book *(continued)*

was pointing down. However, the picture was sufficient to document the behavior, and I can credit *Winter World* for it. Most of my material for book-writing originates from my experiences and from research. It was a thrill to have it the other way round. This time a research publication in an ornithological journal resulted from the book. ∿

Notes

Heinrich, B. and R. Bell. 1995. "Winter Food of a Small Insectivorous Bird, the Golden-crowned Kinglet." *Wilson Bulletin* 107(3):558–561.

Heinrich, B. 2004. "Overnighting of Golden-crowned Kinglets in Winter." *Wilson Bulletin* 115:123–124.

Excerpts from Bernd Heinrich's *Summer World*

As the snow melts and the spring approaches, the animal kingdom awakens. In Summer World, *Bernd Heinrich, bestselling author of* Winter World, *brings us an up-close and personal view of that awakening and rebirth. Almost all life on the surface of the earth derives its energy from the sun, either directly through photosynthesis or indirectly by consuming plants, making summer the time when nature is most active—feeding, fighting, mating, and nesting. From frogs, wasps, and caterpillars to hummingbirds and woodpeckers, Heinrich explores these animals'* ▶

Excerpts from Bernd Heinrich's
Summer World (continued)

adaptations for surviving and procreating during the short window of summer, and he delights in the seemingly infinite feats of animal inventiveness he discovers there. Beautifully illustrated throughout with the author's delicate drawings and infused with his inexhaustible enchantment with nature, Summer World *encourages a sense of wonder and discovery for the natural world and its busiest season. Bernd Heinrich's* Summer World *is available in hardcover from Ecco, an imprint of HarperCollins.*

On Wood Frogs

If animals' main summer preoccupation is a race of reproduction, then the chorusing of wood frogs on a night in early April is the "starting gun." The frogs burst out from under the decaying leaves on the ground and overnight convene onto a just-melted pool, and start their convocation, which is rowdy, loud, and brief. One might presume the males call to attract females specifically to themselves, but now after getting to know them a little better, I think the story of what they do is more interesting. As we shall see, it can involve cannibalism, and more.

On Phoebes

A late March snowstorm earlier in the week dumped inches of snow, but a south wind is melting it fast. A robin sings, and the redwings are yodeling down in the bog. I expect the phoebe back at any time, too. It would be flying north now, aided by a wind at night from Alabama or Georgia, and powering itself to hurry along on the homeward journey back to a mere pinpoint on the continent—the house where I live and from where it had left to go south last September. Such feats of endurance and navigation are routine for many migrant birds, but how they might be accomplished still boggles my imagination, no matter how many "explanations" such as magnetic orientation, use of landmarks, solar orientation and precise timing, use of prevailing winds, are or could be involved. . . .

I woke up in the gray dawn to ▶

the sounds I've long awaited—
a loud, emphatic, endlessly repeated
"dchir*zeep*, dchir*zeep*." His enthusiasm
was infectious. I jumped out of bed
and announced: "The phoebe is
back!" I took a good look at our
friend. There he was perched on a
branch of the sugar maple tree, about
six feet from our bedroom window.
He was dipping his tail up and down,
a phoebe gesture signaling health and
vigor. As I watched this sparrow-sized
bird from up close, I noted his black
cap, white throat bib, and dark
gray back. He stretched a wing and
shook his fluffy plumage, and I felt
transported as if into another being.
I experienced a glow of warmth and
satisfaction, as anyone would when
confronting a marvel of creation that
magically appears on one's doorstep
at almost precisely the time that one
predicted it would come.

On Ruby-throats

A male ruby-throat weighs about
3 grams, only slightly more than a
penny (2.5 grams), and is several
times lighter than many large moth

caterpillars. Its heart and wings beat at 21 and 60 times per second, respectively, while flying north on a journey of about 2,000 miles, then again on the same journey, this time south in the fall. At any given moment it is within hours of starvation, as it needs to consume about twice its body weight per day. No other hummingbird attempts what would appear to be a very risk-prone journey, to a destination often destitute of nectar-bearing flowers.

On Bogs

I brew myself some coffee and then leave to re-visit Huckleberry Bog, a haven for many plants and animals that are not found in the forest. But it is surrounded by forest and edged by a cordon of dense, brushy thickets growing in algal-covered water. I don tough pants and a shirt when forcing my way through, but I don't wear boots—they get filled with ice-cold muddy water when you get to hip-deep holes or beaver channels. Wet, cold feet are obligatory, and I use worn-out running shoes. With ▶

Excerpts from Bernd Heinrich's
Summer World (continued)

some relief I finally break through
to enter the bog proper, where I am
in the open and walk on an ancient
mat of roots and sphagnum (peat)
moss that has grown over a glacial
pond. Some of the same tree species
that grow at the periphery are also
present—red maples, black spruce,
and larch. But here they are scattered
and stunted like bonsais. Like
walking on a waterbed, each footstep
depresses this mat and it sinks a few
inches, to again rebound when I lift
it, hence the name of "floating" bog.
Somewhere still preserved in the
solid bottom below me is the pollen
of the plants that had grown on
surrounding hills after the last ice
age. Did a woolly mammoth or two
break through and leave their bones
here, also? Except for the refrain of
the yellowthroat and six other bird
songs, the bog remains silent. It does
not tell. But a white-throated sparrow
flushes near my feet, and I gaze into
its nest-cup that is sunk into the wet
moss. I admire four blue-green eggs
spotted and blotched in reddish-
brown. What don't I see? Where is

the olive-sided flycatcher? It used to
be always here, perched on the tip
of a tamarack and repeating his
loud clarion call that seemed like
the signature sound of the bog.
Where are the bumblebees . . . ?

In the early morning haze, the
bog's open expanse is a study in rich
greens and pastels. The over-wintered
needles of the stunted black spruce
are bleached to a yellowish tinge
from their previous summer's fresh
blue-green. None of their leaf
buds have yet opened, although the
tamarack, also a conifer, had shed its
golden-yellow needles in the fall and
is now opening all of its buds on its
black and white lichen-encrusted
twigs, revealing tufts of light
bluish-green needles. An intricate,
intertwined tangle of evergreen
perennials—leatherleaf, swamp laurel,
rosemary, Labrador tea, lambkill, and
cranberry—rests on the sphagnum,
and as my feet sink in I see small
clumps of bright pink blossoms of
the swamp laurel, and shining white
ones of the rosemary. At the water
along the edge grow the taller and ▶

deciduous plants—high-bush
blueberries, huckleberry, winterberry,
chokeberry, and privit Andromeda—
all now putting on new yellowish-
to bluish-green leaves. The color
combinations of the leaves, buds,
twigs, flowers, and berries—greens,
browns, yellows, greens, red, gray,
black— are artistically "perfect." I
pick a selection of the plant riches to
take home to make a token watercolor
sketch. It can only be a reminder of
the beauty and perfection of this
place, a "piece of work" of the
Creation. ∾

Have You Read?

THE SNORING BIRD: MY FAMILY'S JOURNEY THROUGH A CENTURY OF BIOLOGY

From Bernd Heinrich comes the remarkable story of his father's life, his family's past, and how the forces of history and nature have shaped his own life. Although Bernd Heinrich's father, Gerd, a devoted naturalist, specialized in wasps, Bernd tried to distance himself from his "old-fashioned" father, becoming a hybrid: a modern, experimental biologist with a naturalist's sensibilities.

Have You Read? *(continued)*

MIND OF THE RAVEN: INVESTIGATIONS AND ADVENTURES WITH WOLF-BIRDS

Heinrich involves us in his quest to get inside the mind of the raven. But as animals can only be spied on by getting quite close, Heinrich adopts ravens, thereby becoming a "raven father," as well as observing them in their natural habitat. He studies their daily routines and, in the process, paints a vivid picture of the ravens' world. At the heart of this book are Heinrich's love and respect for these complex and engaging creatures, and through his keen observation and analysis, we become their intimates too.

Heinrich's passion for ravens has led him around the world in his research. *Mind of the Raven* follows an exotic journey—from New England to Germany, and from Montana to Baffin Island in the high Arctic—offering dazzling accounts of how science works in the field, filtered through the eyes of a passionate observer of nature. Each new discovery and insight into raven behavior is thrilling to read, at once lyrical and scientific.

"Bernd Heinrich is one of the finest living examples of that strange hybrid: the science writer. . . . No definition of God has ever made me feel as comfortable, small, and important in the universe as Heinrich's insight into the mind of the raven."
—*Los Angeles Times Book Review*

WHY WE RUN: A NATURAL HISTORY

In *Why We Run*, Bernd Heinrich explores human evolution by examining the phenomenon of ultraendurance and makes surprising discoveries about the physical, spiritual—and primal—drive to win. At once lyrical and scientific, *Why We Run* shows Heinrich's signature blend of biology, anthropology, psychology, and philosophy, infused with his passion to discover how and why we can achieve superhuman abilities.

"A stunningly original book. It blends personal experience in world class distance running with a firsthand account of the biology of running by one of its leading authorities."

—*E. O. Wilson*

THE GEESE OF BEAVER BOG

With a scientist's training and a nature lover's boundless curiosity and enthusiasm, Bernd Heinrich set out to observe and understand the travails and triumphs of specific individual Canada geese living in the beaver bog adjacent to his rural Vermont home. Heated battles, mysterious nest raids, jealousy over a lover's inattention—all are recounted here in an engaging, anecdotal narrative that sheds light on how geese behave as they do.

Heinrich takes his readers through mud, icy waters, and overgrown sedge hummocks with deft insight, respectful modesty, and infectious good humor, accompanied by his beautiful four-color photographs and the author's trademark sketches.

"Heinrich's lyric writing and attentive observations make goose world come alive. . . . [A] pure joy."
—*Los Angeles Times*

THE TREES IN MY FOREST

Bernd Heinrich takes readers on an eye-opening journey through the hidden life of a forest.

"This lyrical testament to the stunning complexity of the natural world also documents one man's bid to make a difference on his own little patch of land. In 1975 Heinrich . . . bought three hundred acres of logged-over Maine woods and set out

to restore its ecological diversity. . . .
In his ultimate goal of creating a
forest, a place of 'habitat complexity'
vastly different from the sterile
monocultures planted in the name
of sustainable forestry, he succeeds
admirably."

—*Kirkus Reviews*

"These passionate observations of a
place 'where the subtle matters and
the spectacular distracts' superbly
mix memoir and science."

—*New York Times*

Don't miss the next book by your favorite author.

Sign up now for AuthorTracker by visiting www.AuthorTracker.com.